Using your TI-89 in learning electrical circuits
A REFERENCE TOOL BOOK FOR ELECTRICAL ENGINEERING STUDENTS AND PRACTITIONERS
Part I: Linear Resistive Circuits.

Rogelio Palomera-García
U. of Puerto Rico at Mayagüez

Using your TI-89 in learning electrical circuits
Part I: Linear Resistive Circuits.

Copyright © 2016 by Rogelio Palomera-García.
All rights reserved, including the right to reproduce this book or any portion of it in any form.
   TI-89, TI-89 Titanium, TI-92 Plus, and Voyager 2000 are trademarks of Texas Instruments.

Written and designed by:
   Rogelio Palomera-García
   rogelio.palomera.garcia@gmail.com

ISBN-13: 978-1539534372
ISBN-10: 1539534372
Printed by CreateSpace, an Amazon.com Company

Preface

> *The calculator is a tool, not a teacher. If you don't know how to do it by hand, with pencil and paper, then you don't know how to do it with the calculator.*

This is not a textbook on circuit analysis. Neither a manual or a collection of programs for graphing calculators. This is a reference book, and I hope a good complement for your textbook, where you learn to make an efficient use of your TI-89 graphing calculator as an effective tool while studying and applying circuits. In other words, you can focus on the problem solving process while leaving the complicated calculations to the machine. This one will work them faster, specially when it comes to complicated numbers.

The above statement, as obvious as it may look, appears to not be grasped by everyone. These graphing calculators have more computational power than several computers had several years ago. Moreover, too many textbooks make reference to the use of software packages such as MATLAB^{TM} or Mathematica^{TM}, which although are indeed very powerful tools, they are not the kind the student would carry in the pocket. You need at least a laptop, and to invest more money than the cost of one of these calculators to get the same power you need for the applications to circuits that we consider here. This is not to diminish the value and power of MATLAB^{TM} or Mathematica^{TM}; in fact, in my opinion every student should have experience with at least one of them. It is just a reminder that as far as many applications concern, these calculators are easier to use, portable, and cheaper. Let's keep the software packages for more demanding needs. As a bonus, let me tell you that everything you learn in this book is directly transferable, with minor adjustments, to these software tools.

Now, despite all the advantages of the graphing calculators, my experience in teaching circuits and other classes tells me that most students do not benefit of their power. These students spend more than $100 to use the tool exactly as if it was a $10 four-basic-operations pocket calculator. I have nothing against these cheap calculators. When effectively used they are powerful too. My point here is that the real power of the graphing calculator is ignored, and a lot of time is lost in complex calculations. Lost either because the theory behind the calculations is not well understood, or perhaps because the user ignores how to use the calculator, or perhaps because both situations occur.

Although this is not a textbook, I try to provide theory behind the computation. Therefore, we will need some theorems, concepts, and algorithms not usually found in introductory circuit books. The target audience for this book is composed mainly of electrical and computer engineering students, as well as practitioner engineers, specially those preparing the PE exams where the calculators are allowed. Non major engineering students, and engineering technology students can also take advantage of this book. In fact, the use of the calculator in this book is based on theory that is common to many fields of knowledge, not necessarily only circuits.

I stress that the focus of the book is to provide you a tool to be used in the process of learning! As such, programs should be a result of the process, not the main contents of the book. I introduce some programs to illustrate the usefulness of the learned process. But if you do your work, you will find out that you can program easily for many applications

Programs to do somehow limited circuit analyses are available on the internet and may be downloaded. I did some browsing and most of them are quite trivial. You should be able to write better ones soon after you start with this book if you already have some basic programming skills. The dark side of downloading available programs, is that you don't learn. And even worst, you miss the fun!

The book deals only with resistive circuits for two reasons. One, is to keep the length of the book within reasonable limits. The other is to concentrate our efforts on the process and the use of the tool. From the algebraic point of view, most of what is said for real resistive circuits is automatically extended to circuits in the phasor domain using complex numbers. I present an example in the last chapter. Similarly, many techniques in time domain of circuits with reactive elements can be associated to resistive network analysis, as I show with another example.

Anyway, I hope to produce a second part to deal with linear RLC circuits in the time domain and complex phasor domain as well. The phasor domain has many more interesting applications than just transforming real numbers into complex numbers for the analyses. Hence, to include everything in just one volume is an overkill. Both for the reader and for me, the author. ;)

Let me now give you a short description of each chapter. Since this is not a textbook, you are not obliged to read them in sequence. You can start wherever you want, skip sections and at your will. Go back and forth as you prefer; you need to feel comfortable. Of course, most chapters are somehow related, but you can just refresh what you need as you go. On my side, I am free to include or not include specific topics or to follow a specific order. I am not even obliged to include everything that comes in a circuits book. In fact, I have skipped all the definitions and calculations about charge, voltage, energy, power, and current where derivatives and and integrals are used. Good examples can be found at references [Edward05, Voltmer99] mentioned at the end of the book.

Chapter 1 is a brief overview of the views I learned from my professors about what circuit analysis is and how we deal with it. My thoughts sketched in this chapter are reflected in the book organization and layout. Also, a brief review of the different approaches in analysis of circuits is discussed. This chapter could be an appendix, and may be skipped without any loss.

Chapter 2 gives a brief overview of some specific characteristics that we need to know about our tool, at least for our purposes. In this chapter I also cover the conventions on notations so that the rest of the book will not become overload with the explanation step by step of keys to be pressed. The topic of user defined functions is included here. This chapter may be skipped in a first visit, or be used only as a reference.

Chapter 3 gives an overview of lists and matrices, both from the theoretic point of view as well as properties concerning matrices and linear equations. In particular, known theorems are repeated because of the importance they have for circuit analysis. This chapter is also a reference chapter.

Chapter 4 introduces the theorems of source substitution, homogeneity and superposition. I feel these theorems should be introduced early in circuit courses. Many practical algorithms and concepts are based on these theorems. The power of the calculator as used in this book is also based in the application of these theorems and the properties of linear equations solutions. The chapter is of theoretical nature but very practical value. You may skip it, or read it as a reference when needed.

Chapter 5 is a tour through different reduction and transformation methods used in circuit analysis. Series-parallel transformations, delta and wye structures, voltage dividers and some specific applications are introduced here. The chapter has two goals. One, to stress the need to think on planning that intuitive path you normally follow when working with pencil and paper to obtain real benefit of the tool. Second, to show clearly how the features of your calculator may be exploited when these methods are used. Some user defined functions are introduced.

Chapters 6 and 7 work the traditional nodal and loop analyses, respectively. Each chapter starts with the traditional constraints of using only one type of sources. But the inclusion of both voltage and current sources in any type of analysis is introduced too. These are the so called the modified nodal and loop analyses. Programming for nodal analysis and the particular case of mesh analysis are introduced in these chapters, with examples. The chapter on loop analysis is shorter than that of nodal analysis because node based description is more direct and based only on the topology of the circuit. In fact, nodal analysis is the basis for the popular SPICE simulator.

Chapter 8 to *Chapter 10* illustrate the power of the tool beyond writing and solving equations, when you combine theory and algorithms. Let me say that these chapters were the real motivation for me to write this book, since it was suggested by colleagues who saw me using the calculator in circuit calculations. These circuits cover transfer functions, Thevenin and Norton representations, and two-ports.

Chapter 11 is an appetizer with a short tour beyond the scope of this book, applying what has been learned in other situations. It may also serve as a motivation for the companion books. For me, to finish writing them. For you, to read them if they are published.

Some techniques included in the book were developed by myself, without claiming with this statement that I was the first to use them. I just have not been able to find those methods elsewhere. The references at the end of the book are a short non ordered collection of books where several ideas were extracted or methods rediscovered. I apologize to the many authors not included. I hope these references can be a source of inspiration for you as they were for me.

This book would have not been possible without the help and encouragement I received from several persons. First of all, I say thanks to those colleagues who in the mid 90's suggested I should write the book. I did it very slowly. In fact, the original unpublished manuscript was written for the TI-85! A second unpublished manuscript with a more general scope was circulated among few people around 2005. I thank Prof. Baldomero Llorens, from the U. of Puerto Rico at Mayaguez, for his criticism pointing out the main flaw of the original manuscript, which made it quite unattractive and dull. His comments made me change the perspective for this book completely.

Prof. Edwin Delgado, from the Interamerican University at Aguadilla, is responsible with his comments for the introduction of chapter 4, as well as the reorganization of several sections in chapter 5, and the rewriting of some material in chapter 6. Thank you for his accurate and to the point criticisms to the manuscript.

My special gratitude goes also to my friend Pedro Escalona. When he was my student, he read the original manuscript and kept a copy of it long after I had lost mine. Thanks for his encouragement to start again and for lending me his copy so I could review my original approach when I finally decided to try again. My original computer files were corrupted and I did not have all the planning available. Now working as a Texas Instruments engineer, he has taken his time to help me take the manuscript to the last step.

My students Jeshua Colón, Luis Millán, Carlos Rivera, and Santos García tested some sections and examples of the book in their calculators. The fact that not all the calculators used were TI brand reinforces the original hypothesis that the methods are, as one should expect, independent from the calculator. By going through text and examples, they corrected several typos, and put order in many messy paragraphs. I initially tried to include several calculators in the book, but some testing with other students convinced me to concentrate in only one model.

Many people from Texas Instruments encouraged me to go on and finish the book; not all of them still work there, but they were at the moment. Among the TI'ers, past and present, let me mention Gregg Lowe, Art George, Praful Madhani, Juan Guerra, Modesto García, and Charles Parkhurst. My sincere thanks to all of them and many others not included in this short list.

Manuel Jiménez, professor at the U. of Puerto Rico at Mayaguez, also provided me with some ideas to make the book readable, and hints on the cover design. Last, by not least, my wife who was patient enough to see me deleting and rewriting pages along all these years, still encouraging me to go on.

All errors, grammar or technical, are my responsibility. I thank in advance for any error that you find and let me know. By experience, I know that errors are somehow analog to lies: if you repeat them constantly, you finish believing they are correct. You may contact me to leave your comments, criticism and let me know your thoughts about this book. Just write me to r.palomera-garcia@ieee.org

Perhaps in the near future I will open a related website for using of the calculator in other ways too. But not right now! I don't promise that.

I hope you enjoy the book and find it useful! I appreciate and thank you in advance for comments, criticism and suggestion you may have.

Rogelo Palomera-Garcia
r.palomera-garcia@ieee.org

Contents

Preface	i
1 Preliminary considerations	**1**
1.1 Physical vs. Model Circuits	1
1.2 Our focus: linear resistive circuits	2
1.2.1 Using this book for AC circuits	4
1.2.2 Circuit equations and solving methods	4
1.2.3 Hand analysis vs analysis with calculator	5
1.2.4 Programming and using programs	5
2 Practical Matters about the calculators	**6**
2.1 Texas Instruments calculators used in this book	6
2.2 Keys and key typing	6
2.2.1 Cursor keys	6
2.2.2 Key values	7
2.2.3 Notation for key entries	8
2.2.4 Powers of 10 using EE	8
2.2.5 Notation for entries on command line and command results	8
2.3 Setting modes	10
2.3.1 Numeric Display Formats:	10
"EXACT/APPROX" menu	10
"Display digits" submenu	11
Exponential Format submenu	11
2.3.2 Angle mode	12
2.3.3 Real/Complex mode	12
2.4 Variables and Storage	13
2.4.1 General remarks: names	13
2.4.2 Storing values in variables	13
2.4.3 Entry function	14
2.4.4 Variable ANS	14
Recalling previous answers or entries from stack	16
2.4.5 Using several previous answers or entries in calculations	16

2.5		Programming and User Defined Functions	16
2.6		A program as a user defined function	17
2.7		User defined functions	17
	2.7.1	Algebraic non-parameter functions	18
	2.7.2	Functions with parameter	19
2.8		Custom Menus	20

3 Lists and Matrices   21

3.1		Lists	21
	3.1.1	Lists and your calculator	22
		List variables	22
		Retrieving and displaying elements of a list:	22
		Editing the list	22
		Augmenting the list	22
	3.1.2	List functions and list menus	22
	3.1.3	Working with Lists	23
3.2		Matrices	25
3.3		Basic definitions and operations	25
	3.3.1	Some basic definitions	26
	3.3.2	Submatrices	26
	3.3.3	Partitioning of matrices	27
	3.3.4	One remark on multiplication of matrices	27
	3.3.5	Matrices and linear combinations	28
3.4		Matrices and Calculators	29
	3.4.1	Matrix menu	29
	3.4.2	Matrix variables and operations	29
		Dot operations in the TI-89	29
	3.4.3	Creating matrices	30
		Using the calculator editor tool	30
		On the command line	30
	3.4.4	Retrieving and editing elements	31
	3.4.5	Retrieving and editing rows and columns	31
	3.4.6	Submatrices	32
		Deleting rows and columns	32
		The `SubMat` function	32
	3.4.7	Matrix functions in the matrix menu	33
		Using `augment(` function	34
3.5		Matrices and Linear equations	35
	3.5.1	Solving linear equations	35
	3.5.2	Multiple systems and knowns. Matrix **B** instead of vector **b**	37
	3.5.3	Exchanging knowns and unknowns	38

4 Four Network Theorems and Applications — 40
- 4.1 Substitution Theorems — 40
- 4.2 Homogeneity and Proportionality — 43
- 4.3 Superposition Theorem — 44
- 4.4 Closing remarks — 48

5 Analysis by transformations and reduction — 49
- 5.1 Series and Parallel Connections — 49
  - 5.1.1 Equivalent resistance and conductance formulas — 50
  - 5.1.2 Another example of Series-Parallel Reduction — 52
  - 5.1.3 Using Lists in series-parallel — 54
  - 5.1.4 Function Parallel pl(z) — 56
    - An example with Homogeneity property — 60
- 5.2 Delta-Wye and Wye-Delta transformations — 61
  - 5.2.1 Working formulas directly — 62
  - 5.2.2 Using functions — 63
  - 5.2.3 Programming the transformations — 66
- 5.3 Voltage and Current Dividers — 67
  - 5.3.1 Two-resistors dividers again: using conductances — 71
    - Loaded Divider — 71
    - Passive adder with voltage divider: applying superposition — 73
  - 5.3.2 Two-resistors divider and it's Thevenin equivalent — 77
- 5.4 Source transformations — 78
  - 5.4.1 Shifting theorems, and extended source transformations — 81
- 5.5 Some Basic Operational Amplifier structures — 82
- 5.6 Closing Remarks — 85

6 Nodal Analysis — 86
- 6.1 Introduction to the theory of nodal analysis — 86
- 6.2 Resistances and current sources only circuits — 88
  - 6.2.1 Theoretical principles — 89
    - Component of G_{mj} of Y_{mj} due to resistances. — 90
    - Voltage-controlled current sources — 91
  - 6.2.2 Rule to generate the equations — 92
  - 6.2.3 Scaling units — 98
  - 6.2.4 An example with dependent sources — 99
  - 6.2.5 Modifying a circuit — 100
  - 6.2.6 Circuits with voltage sources using source transformation — 102
- 6.3 Other considerations for nodal equations — 103
  - 6.3.1 Indefinite admittance matrix — 103
  - 6.3.2 Symbolic sources, time function and superposition — 103
  - 6.3.3 Programming nodal equations I — 105
    - Preparing the inputs — 106
    - Programming the calculator — 108
- 6.4 Circuits with voltage sources and other elements — 109

CONTENTS

    6.5 Modified Nodal Analysis (MNA) 111
              Working MNA with known potentials 116
        6.5.1 Programming Modified Nodal Analysis 117
              Circuit Description 118
              Program pseudocode for mna(n,r,is,gs,vs,ks,ccs) 120
              Developing the program for mna 122
    6.6 "Reducing" number of equations 124
        6.6.1 Eliminating all unknown currents by inspection: Supernodes 124
              Applying superposition 127
        6.6.2 "Hybrid" reduced MNA 128
        6.6.3 Working with Operational Amplifiers 129

7 Loop Analysis 132
    7.1 Loop currents and loops selection 132
    7.2 Circuits with resistances and voltage sources 135
        7.2.1 Circuits with only resistances and independent voltage sources 135
              Scaling units 142
        7.2.2 Current-controlled voltage sources 142
        7.2.3 Indefinite mesh matrix 145
    7.3 Programming mesh equations I 146
        7.3.1 Pseudocode for program 146
        7.3.2 Preparing input for the program 147
        7.3.3 Program for the calculators 148
    7.4 Loop Analysis including Current Sources 149
        7.4.1 Modified Loop Analysis 150
        7.4.2 Reducing the number of equations 152
    7.5 Further considerations 155

8 Network Functions 156
    8.1 Definition of the network functions 156
    8.2 Finding the network functions 158
        8.2.1 Principles of calculation 158
        8.2.2 Calculating functions: examples 160
    8.3 Open and short circuit transfer functions 164
    8.4 Chapter summary 166

9 Superposition: A powerful tool 167
    9.1 Thevenin and Network Equivalent Circuits 167
        9.1.1 Basic theory 167
        9.1.2 Finding equivalents: principles 168
        9.1.3 Finding equivalents: examples 169
    9.2 Calculating two-port parameters 173
    9.3 Finding two-port parameters from equations 174
    9.4 Summary 178

10 Two-port and three-terminal networks — 179
- 10.1 Basic Definitions 179
- 10.2 Two port parameters 180
  - 10.2.1 Definition of Parameters 180
  - 10.2.2 Transforming parameters 180
    - Simple cases 182
    - Conversion by setting up and solving the equations 182
    - Programming transformations 183
- 10.3 Applying two-port parameters 187
  - 10.3.1 Terminated two-ports 187
  - 10.3.2 Tandem Connection 190

11 What's Next — 194
- 11.1 Complex circuits in steady state domain 194
  - 11.1.1 Elementary notation and display for complex numbers 195
  - 11.1.2 Complex numbers and calculator 196
    - Lists and Matrices: 196
  - 11.1.3 An Example 196
- 11.2 Short reference to time domain circuits 198
- 11.3 Final remarks 200

12 References — 201

CHAPTER 1

Preliminary considerations

This chapter offers a brief introduction to the topic of circuit analysis so the organization and procedures in the rest of the book are better understood. The reader may skip the chapter without loss in calculator skills.

1.1 Physical vs. Model Circuits

A *physical circuit* is built with the interconnection of *physical elements or devices*. The size of the circuit may be in the nanometers or micrometers range up to hundreds of meters. The number of devices may go from just two of them, to trillions of elements. In applications of physical electrical circuits we are concerned with the interaction of electrical and magnetic magnitudes such as currents, voltage, electric charge, electric fields, magnetic fields, power, energy and so on.

Unfortunately, it is impossible to fully describe a physical circuit in mathematical terms, or for that matter, even in physical terms. This is true also for simple circuits consisting of just two elements! To deal with physical systems, it is necessary to develop models which are limited descriptions of the devices and their interconnections, focused on some particular characteristics. The validity of an equation and the result or calculation are limited by the accuracy of the models used.

Within an interval of frequencies that goes from DC (0 Hz) up to an upper bound defined by the physical circuit, a device is modeled by equations referred to as *element equations* which ignore the size and shape of the device. On the other hand, the interconnection obeys to a separate set of equations derived from *Kirchhof's Voltage Law* and *Kirchhoff's Current Law*, which do not depend on devices. The result of this process of modeling, is what we call *model circuits*.

How this process fits in the world of real world applications is illustrated in Fig. 1.1 on the following page. We start from a physical circuit from which, based on the Laws of Physics, physical elements behavior and many other factors to consider, we go through a modeling process and arrive at a model circuit, composed of ideal elements. In this circuit we apply the methods learned in circuit courses, some of which are reviewed in this book, to arrive at some results. These have yet to be interpreted so the physical process can be predicted within a certain accuracy. This prediction still needs to be validated by physical experimentation. If validation fails, the process is to be reviewed first by checking the analysis methods, and the final interpretation. If an acceptable result is still not present, then the modeling procedures and the initial model circuit are to be reviewed and changed if necessary.

Using this figure as a reference, you should know that this book, as well as most material in your circuit course, is mainly focused on the methods used to go from the initial model to the result model. In other words, *we work our engineering process with the model circuit, not the physical circuit.* The best proof we have that this methodology is successful is technology working as we know it. In fact, the calculator you are using was designed in this way! For that reason, hereafter, the term "circuit" always refers to model circuit.

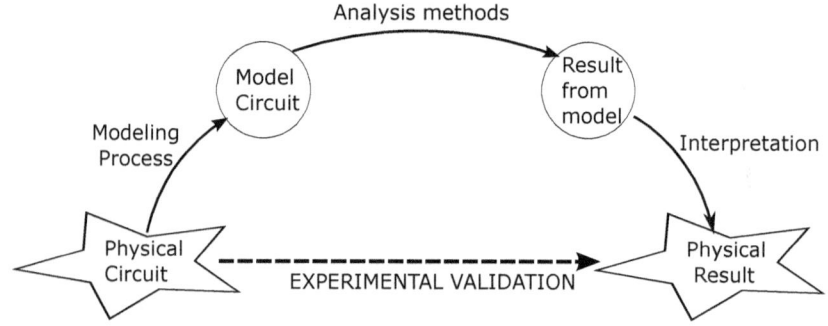

Figure 1.1: Rough description of circuit modeling and analysis.

1.2 Our focus: linear resistive circuits

Let us circumscribe the focus of this book. We deal only with analysis of linear resistive circuits. By analysis it is meant to solve a set of circuit equations by any of the methods available. Sometimes our problem requires only some specific quantities, and we stop at this requirement. But it is important to understand that no matter the method, when it is necessary we should always be able to find all voltages and currents in the circuit.

To illustrate this remark, the set of equations can be the ones from the nodal method, which solve the potentials of nodes and eventually some other currents. Well, every other current and voltage not appearing directly in the set of equations must be a function of these potential and current variables of our equations.

Considering another example, when we proceed by circuit reduction, we should be able to go backwards in the process to find the voltages and currents.

1.2. OUR FOCUS: LINEAR RESISTIVE CIRCUITS

In our circuit courses, we learn how to use the voltage and current in elements to find other magnitudes such as transferred electric charge, absorbed or delivered power or energy, and so on. We will make use of these concepts now and then to illustrate our use of calculator in this process. But the book is focused on circuit analysis within the limits of the above description.

In this book we furthermore limit ourselves to linear resistive circuits with specific elements. Namely, circuits are composed by resistances, linear dependent sources, independent sources, and ideal operational amplifiers. The symbols and equations for these elements are shown in Table 1.1.

Table 1.1: Linear Resistive Elements

Element	Symbol	Equation
Resistance		$V = RI$ or $I = GV$
Independent Voltage source		V given
Independent Current source		I given
Voltage Controlled Voltage source (VCVS)		$V = kV_x$
Voltage Controlled Current source (VCCS)		$I = gV_x$
Current controlled Voltage source (CCVS)		$V = rI_x$
Current controlled Current source (CCCS)		$I = \alpha I_x$
Ideal Operational Amplifier (OA)		$V = 0$ $I_1 = I_2 = 0$

Voltage is measured in Volts (V), current in Ampere (A), Resistance R and transresistance r in Ohm (Ω), Conductance G=1/R and transconductance g in Siemens (S) or ($\Omega^{-1} = 1/\Omega$).

In sources, the voltage and current at the element are unrelated. The undefined magnitude is determined by its compliance to the respective Kirchhoff's law. In dependent sources, however, the controlling magnitude may be the one in the element, but this is an exception, except for specific applications. In an ideal operational amplifier, the output current and voltage are not related to the input magnitudes, but are determined by the equations of connection.

1.2.1 Using this book for AC circuits

Circuits in the phasor domain are not considered in this book. About these circuits, two remarks are worth pointing out.

The first one is of outmost importance, because it makes this book very useful for ac circuts. Namely, the algebraic procedures to find all the phasor voltages and currents in the circuit follow exactly the same methods and procedures that this book covers. An example is shown in the final chapter. With the exception of some properties for resistive circuits, we can affirm that everything presented in the book is applicable to phasor circuits.

The second remark justifies this being the first part of our study. Namely, there are theoretical and practical considerations worth dealing with separately. Also, applications to power systems and circuits in ac domain deserves a treatment by itself.

I invite the reader to take a circuit with complex impedances and phasors for currents and voltages, and apply the methods of reduction, nodal, loop, two ports and others to the circuit. Again, read section 11.1.2 and look at the example there to see what I mean.

1.2.2 Circuit equations and solving methods

At the end, the target in circuit analysis is to solve the circuit equations. The *original* set for a circuit of B elements consists in 2B equations, of which B are derived from Kirchhoff's laws, and the other B equations correspond to elements. To tell you the truth, these are too many equations!

To simplify this task, different methods have been developed. These can be divided in two great groups: a) working by reduction and transformation, and b) setting up a reduced number of equations following some rules.

In reduction and transformation methods, sub circuits are identified and substituted by simpler or smaller equivalent configurations, perhaps one element, in such a way that the new circuit is smaller or easier to deal with. We could say that these methods are akin to reducing the number of equations and variables. The procedures in this group depend on your ability at identifying structures and applying your intuition. They help you to develop, on the other hand, your intuition for easier recognition of properties and structures, and for design purposes. Some structures are so common that they are worth programming.

The other group consists in using systematic methods of establishing sets of equations, which are in a smaller number than the 2B original equations. The most

1.2. OUR FOCUS: LINEAR RESISTIVE CIRCUITS

popular among these methods are the nodal and loop analysis methods. These are the ones usually taught in undergraduate courses and the ones we deal with in this group. There are other methods that the reader may consult elsewhere.

The methods in the second group are more suitable for mathematical algorithms, because they follow well established step by step processes. As a consequence, they are easy to program.

1.2.3 Hand analysis vs analysis with calculator

Most students I know use the calculator as an extension of their pencil, mainly to execute the operations. Before following this path, it is worth looking at what your calculator has and how working with it will be different.

When working with pencil and paper, you try to use simpler formulas and simplify calculations before proceeding further. This method is not necessarily the best when using calculators. For example, when combining three or more resistances in parallel by hand, it's easier to work in steps of two resistances per round. Why? Because multiplying R1 and R2, and then dividing by (R1+R2) is more intuitive than adding the inverses $1/a$ and $1/b$, and then taking the inverse again. In fact, I've found "programs" in internet that work in this way for three or more resistances.

But in the calculator it is easier to work $(1/a + 1/b + 1/c)\wedge$-1 directly! One formula for any number of resistances in parallel.

This is just one of many examples that tell us that you must do your homework in setting your mindset for calculator use. Know your calculator and your theory, and look for methods that let you apply your calculator effectively. This principle applies to any technological tool, let it be a calculator, a spreadsheet, software like Matlab^{TM}or Mathematica^{TM}, and so on.

1.2.4 Programming and using programs

I repeat what I already said before: " if you don't know how to do it with pencil and paper, you don't know how to do it with the calculator". Running a program doesn't mean you work by yourself the task the program is doing.

To illustrate my point, if you enter the data of the circuit to the SPICE simulator, you will get the node potentials. You did not calculate the node potentials, and SPICE did not learn you how to do it. If you don't know how to obtain those potentials, it is not by using SPICE that you will learn how to do it.

My first advice here is then: do not attempt to write down a program in your calculator until you are sure you have learned yourself how to solve the problem. Of course, once you know how to do things, write the program and use it! You have earned that privilege! Or you may download one from the internet for that matter, or from your friend's collection. On the other hand, planning and writing a program may be an adventure by itself, because at the end you will follow paths suitable for the calculator that would not be used with pencil and paper. It is worth the experience. And the prize is very rewarding.

CHAPTER 2

Practical Matters about the calculators

In this chapter I explain several points related to the notation used to simplify writing for me, and reading for you, as well as remarks relating to features of the calculator that will allow us to make better use of them. You do not need to read this chapter first if you have already a good understanding of your model. Also, you may refer to the chapter whenever you need something in particular.

2.1 Texas Instruments calculators used in this book

This book does not intend to be a handbook for the the TI-89 , illustrated in Fig. 2.1. The reader may consult the guidebooks available online at

https://education.ti.com/en/us/products#product=graphing-calculators

for more information. For our purposes, the TI-89 family includes the TI-89, TI-89 Platinum, Voyage^{TM} 200 and the no longer available TI-89 and TI-92 Plus (but still in use). Also, many things explained here also apply to TI-86 and TI-85.

Let us now look at some features that we may need.

2.2 Keys and key typing

Let me introduced now some remarks on notation for keys.

2.2.1 Cursor keys

The cursor in the graphing calculators may be moved along up, down, left or right directions using the cursor keys. These are usually arrows or triangles located on the key pad such as ◄, ►, ▲ and ▼. The cursor itself appears blinking or else highlights the word or entry selected.

2.2. KEYS AND KEY TYPING   7

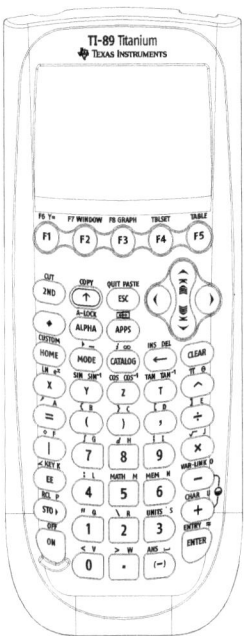

Figure 2.1: TI89 Key distribution (*Image Courtesy of Texas Instruments, Inc.*)

I used the triangles alone, to indicate pressing of a cursor key.

2.2.2 Key values

Almost all keys on the pad have more than one value. The face, or primary value, which is the one printed on the key itself, and alternate, or secondary, values shown above the key. The secondary values are obtained pressing an auxiliary key before the key itself.

The TI-89-Platinum has the auxiliary keys $\boxed{\text{2nd}}$, $\boxed{\text{ALPHA}}$, and the diamond key $\boxed{\blacklozenge}$. The auxiliary keys are colored in such a way that they match the color of the secondary value associated with it. The reader may check the image in color at the front cover.

Letters Letters in the TI-89 and TI-89 platinum are obtained using the $\boxed{\text{ALPHA}}$ auxiliary key. The TI-89 in particular has the letters X, W, Z and T as primary values of the respective keys, so no auxiliary key is needed here. On the other hand, the TI-92 Plus and the Voyage^{TM} 200 models have a QWERTY keyboard, so a letter can be directly typed pressing the respective key.

Function keys Function keys $\boxed{\text{F1}}$, $\boxed{\text{F2}}$ and so on are available for opening menus and other features. Please refer to your manuals.

2.2.3 Notation for key entries

Sometimes I will make reference on how a particular entry was typed on the command line.

1. Framed characters of any type are keys to be pressed. All framed entries have the face value of the key inside the frame, and not what they represent.

   - Thus $\boxed{=}$ stands for the the key with "=" on it. In the TI-89, the sequence $\boxed{\text{ALPHA}}$ $\boxed{=}$ produces the letter "a".

2. Characters to appear on the entry line, or stack, are written in `typewriter font` without frame, (for example, 2 instead of $\boxed{2}$), unless there may be confusion. In the two examples shown below the key $\boxed{\text{ENTER}}$ is included to avoid confusion with the sequence `e n t e r` and $\boxed{(-)}$ is shown explicitely to avoid confusion with (-).

   - 3 + 2 / 5 $\boxed{\text{ENTER}}$ corresponds to the sequence $\boxed{3}$ $\boxed{+}$ $\boxed{2}$ $\boxed{\div}$ $\boxed{5}$ $\boxed{\text{ENTER}}$. $\boxed{\div}$ produces the slash /, and sometimes I write the slash and others the framed entry.

3. Letters to be entered may be shown in capital or small case, although display on the calculator is always in small case unless shift is used. For cases in which letters require the modifying key $\boxed{\text{ALPHA}}$, the steps will not be included in the expression unless some confusion may arise.

   - Thus I write 2 A + 3 instead of 2 $\boxed{\text{ALPHA}}$ $\boxed{=}$ + 3

Table 2.1 shows the sequence for several characters used in the book. The modifying or auxiliary keys are also used to open menus that we will need, in particular for matrices and lists. They will be mentioned later.

2.2.4 Powers of 10 using EE

When you want to enter a datum that includes a power of 10, such as 1.2345×10^4, 3567.5×10^{-6}, you may use the $\boxed{\text{EE}}$ key. For example, the previous data may be entered as follows:

1 . 2 3 4 5 $\boxed{\text{EE}}$ 4   and   3 5 6 7 . 5 $\boxed{\text{EE}}$ $\boxed{(-)}$ 6

On the result window, the exponent appears with the letter E, as in `1.2345E4`

2.2.5 Notation for entries on command line and command results

Let me introduce the notation used in the book regarding the inputs on the command line, and what results from the input.

2.2. KEYS AND KEY TYPING

Table 2.1: Some characters using a modifying key

Character	Sequene
{	[2nd] [(]
}	[2nd] [)]
[	[2nd] [,]
]	[2nd] [÷]
i	[2nd] [CATALOG]
° [3]	[2nd] [∣]
:	[2nd] [4]
;	[2nd] [4]
" (quotation mark)	[2nd] [1]

[1] Used to implement degree units for angles during entry, overriding angle settings.

To indicate the result going onto stack from an entry line after pressing [ENTER], I use a right arrow, →. Thus, the next expression means that the result 8 results after writing 3+5 on the command entry line and pressing the key [ENTER].

3 + 5 [ENTER] → 8

Sometimes the key [ENTER] may be omitted, as shown next, if no ambiguity arises.

3 + 5 → 8

On the other hand, when a function or program is called and the name appears on the command line after a sequence of keys, I use a ⇒ to show how the procedure As an example, conside the predefined list function notation sum(, which may be called with the sequence [2nd] 5 3 6. With this in mind, the following line shows the sequence of keys to get the input on the entry line sum({3, 6})

[2nd] 5 3 6 { 3, 6 }) ⇒ sum({3, 6})

The first two are used to open the MATH menu, as a secondary function of the key [5], so we can further simplify notation as

[MATH] 3 6 { 3, 6 }) ⇒ sum({3, 6})

By the way, predefined or user defined functions or programs may be also be invoked by directly writing the name on the command line line. Thus typing [MATH] 3 6 is equivalent to typing s u m (

2.3 Setting modes

Now, let us look at the mode settings in the calculator which determine how data results are displayed and, in some cases, also read from the entry line. For our purposes, we concentrate in the settings for number display, angle units, and for complex data display. Most settings, not all, affect only the display of the answer, not the input format of data.

To define your settings, open the mode settings window by pressing [MODE]. Once all the modes have been set, a final [ENTER] takes you back to the HOME environment.

Once the MODE window is open, use the up and down cursor keys to navigate through it and choose the group of settings to be defined. Then the right and left cursor keys to select the desired mode. When your selected mode is highlighted, press [ENTER].

In the TI-89 there is a special submenu called *Apps Desktop*. In this submenu I prefer the option "OFF" because it is faster for me to open an application. But this is my preference and it makes no difference as far as you feel comfortable with your option.

2.3.1 Numeric Display Formats:

There are three characteristics to be set relative to the display of the numbers in the result: a) If in exact or approximate form; b) the number of digits as well as the number of digits to the right of the decimal point; and c) the so called exponential or numeric format.

"EXACT/APPROX" menu

The submenu and has three options: *Auto, Exact, Approximate*, described next.

Exact Any result that is not a whole number is displayed in a fractional or symbolic form ($1/2$, $3\sqrt{2}$ etc.), assuming no decimal point was typed.

Approximate or Dec All numeric results, where possible, are displayed in floating-point(decimal) form. That is, either an integer or a number containing a decimal point.

Auto Provides the EXACT form where possible, but the APPROXIMATE form if the entry contains a decimal point.

In most cases, I prefer the *APPROXIMATE* feature, and this will be default in this book. Whenever the result is not exact, the least significant digit is always rounded.

2.3. SETTING MODES

"Display digits" submenu

You can control the number of digits to be displayed, or else to be displayed in the decimal part, using the "Display Digits" submenu options. The option *FLOAT* alone shows the total number of digits displayed and the digits in the decimal part varies, depending on the result. Now, for the other options:

Float N: With *FLOAT N*, where integer $N \geq 1$, will display results rounded to a maximum total number of N digits. If necessary, will automatically switch to scientific mode to maintain the N digits.

Fix N The option *FIX N*, $N \geq 0$, will determine the number of digits in the decimal part. Notice that if you select FIX format, then extra 0's may be added to the decimal part if necessary.

Exponential Format submenu

The exponential format deserves particular attention because when used wisely our results are easier to read and faster to interpret. The EXPONENTIAL mode has the following options. In all cases the number of digits are displayed according to the settings described before.

Normal: result shown in full. When not exact, the least significant digit is rounded.

Scientific: Result is displayed in the form `n.ddddEM`, where $0 < |n| < 10$. This corresponds to the normal scientific notation $n.dddd\ldots \times 10^M$. For example, `2.4567E-2` corresponds to 2.4567×10^{-2}. Beyond certain limits, normal displays switches to the scientific format automatically.

Engineering: Similar to the scientific format, except that $0 < |n| < 1000$ and M is always a multiple of 3.

The above descriptions stand for the displayed results, not for the input. You may enter the numbers in any format you want. If we are working in scientific mode display, datum 2.4567×10^{-2} may be entered as `2.4567` EE `(-)` `2`, `0.024567`, `245.67` EE `(-)` `4`, or any other form. In all cases the displayed result will be the same.

One advantage of the engineering notation is that the powers of ten are directly related to the prefixes used in engineering and shown in Table 2.2.

Prefixes may be adapted to situations where units are already in a scaled value. For example, talking of current in Amperes, 2E-3 means 2 mA. Yet, if the current in calculation is already in μA units, then 2E-3 will mean 2 nA.

This feature makes it easier to do calculations and interpret results without the need to enter powers of ten. We will have opportunities to illustrate this remark.

Table 2.2: Prefixes for the SI units

Prefix	symbol	Power of 10	Display
femto	f	10^{-15}	E-15
pico	p	10^{-12}	E-12
nano	n	10^{-9}	E-9
micro	μ	10^{-6}	E-6
mili	m	10^{-3}	E-3
N/A		10^{0}	E0
kilo	k	10^{3}	E3
mega	M	10^{6}	E6
giga	G	10^{9}	E9
tera	T	10^{12}	E12

2.3.2 Angle mode

This mode sets the units in which angle values are interpreted when input and displayed. Three options are offered: *RADIAN (Rad), DEGREE (Deg) or GRADIAN*. If you work with trigonometric functions or complex numbers, this is an important setting to define. For inputs, the setting may be overridden with the $^\circ$ feature (see Table 2.1) if you want to enter angles in degrees, and ^{r} if the angle is in radians. For example,

sin(30) = 0.5 in degree mode, sin(30) = -0.988 in radian mode, and sin(30) = 0.454 in gradian mode, but sin(30°) = 0.5 in any mode.

2.3.3 Real/Complex mode

When in *REAL* mode, complex results are not displayed unless your input has a complex entry.

When in *Rectangular* mode, complex results are displayed in rectangular format $a + bi$

When in *Polar* mode, complex results are displayed in polar format $a\angle\phi$ when angle mode is in degrees, and in the polar format $e^{i\phi} \cdot a$ when in radian mode.

Regarding the input of an angle ϕ, the calculator will use the angle settings being used.

For the input of complex numbers, you may use any format and the result will be displayed according to your settings. An input in polar form using the $\angle$ symbol must be enclosed in parenthesis, $(A\angle\phi)$. When the angle settings is in radians, you have the option of using the exponential form for the input, $Ae^{i\phi}$. You can mix numbers in different formats. For example, assuming radian mode for the angle and rectangular format for display:

((3+2i)* (4 $\angle$ .38))/(6*e$\wedge$(.56i) + 1.4 - 0.79 i $\rightarrow$ e$^{.6139 i}$ 2.086

2.4. VARIABLES AND STORAGE

If you try the same input in degree mode, the calculator will show an error because of the invalid exponential polar input used.

2.4 Variables and Storage

Variables are very convenient to work with when using calculators. The tools for dealing with memory and variable management are found in the guidebook. Here, I introduce several points to remember and understand about variables. In the calculators used, the names are valid for numbers, lists, matrices, programs, and functions

2.4.1 General remarks: names

- You must give variable a name before using it. If the variable does not already exist, the calculator will create it.

- If the variable already exists, and you assign a new value to it, the calculator updates it automatically.

- Variable names start with a letter. If it starts with a number, the calculator will assume the operation of multiplication. Thus "2R" is interpreted as "2×R".

- If no numerical value or algebraic expression has been assigned to a variable, the TI-89 deals with it as an algebraic variable. You may check if there is a value assigned to a value by typing the variable on the entry line and looking at the result.

    – Example: if no value has been assigned to a X, then
    x → x
    x^2 + 9 → x^2 + 9

- Several names are reserved in the TI-89 . Among these we have **y1, y2** ,... **r1, r2,** .... When attempting to define one of them, an error message will be displayed

- If a variable name has been associated to a function or a program, it becomes a reserved one. The only way to free it from being reserved is to either change the name or delete it. Consult the manual for details.

2.4.2 Storing values in variables

Let us now look at the process of storing values in variables. There are two ways of doing it

1. Using the store (*STO*) function with the key $\boxed{\text{STO}\blacktriangleright}$ in the format
Value or expression $\boxed{\text{STO}\blacktriangleright}$ *Variable*

- Example: to assign the value 4 to variable X, 4 [STO▶] X → 4

2. Using the **Define** command in the format *Define variable = value or expression*

    In the TI-89, the Define command is obtained with the sequence [F4] 1:
    [F4] 1 X = 4 ⇒ Define X = 4

An expression with an undefined variable, say X, may be assigned to another variable, which becomes then a function of the first one. This is discussed in subsection 2.7.1 on page 18.

2.4.3 Entry function

The calculator may recall up to 99 expressions that have been entered. One way is to move the cursor key until the desired entry on stack is highlighted, then press [ENTER] to paste the entry on the command line.

The sequence [2nd][ENTER] executes the **entry** function, which consists in pasting on the command line the last entry recorded on the stack. If you repeat the sequence, you will navigate through previous entries. The entry is not executed immediately, so you may edit it.

Pressing [ENTER] after a sequence, does not erase the sequence from the command entry line. Hence, pressing [ENTER] alone will repeat the same command. This is illustrated in the example given in next section.

2.4.4 Variable ANS

A secondary value of the key [(-)] is the variable ANS. This variable stores the last result and is updated after each transaction.

In fact, ANS is an array variable, so when you press [2nd] [(-)] you get ANS(1) on the command entry line, which stands for the last result calculated. Previous results are also available as array variables ANS(2), ANS(3), etc. After each transaction, these values are updated. This particular feature of the TI-89 becomes handy sometimes, as illustrated in the next subsection and examples in later chapters. The index 2, 3 and so on must be edited manually. Thus, to get ANS(2) you may type [2nd] [(-)] [←] [←] 2). One of many advantages of ANS is in reducing rounding errors.

This variable has an important characteristic. If you start the command line directly with an operation, the assumed operand is ANS. The following lines illustrate this comment. Each line shows first what you enter on the entry line, and to the right what it actually appears on the line. Some lines would not apply to your calculator because of the lack of the key:

+ 2 ⇒ ans(1) + 2

2.4. VARIABLES AND STORAGE

$\boxed{\wedge}$ 2 $\Rightarrow$ ans(1)^2

$\boxed{\div}$ 2 $\Rightarrow$ ans(1) / 2

Let us now illustrate an example using the ANS variable.

Example 2.1 *The objective is to calculate the expression*

$$M = \frac{1.26 \times 10^3 (287 + 33.25 \times 10^2)}{1.26 \times 10^3 + 287 + 33.25 \times 10^2}$$

Settings are in Engineering mode. The sequence is as follows:

287 + 33.25 $\boxed{\text{EE}}$ 2 $\boxed{\text{ENTER}}$ $\rightarrow$ 3.612E3

$\boxed{\times}$ 1.26 $\boxed{\text{EE}}$ 3 $\boxed{\div}$ (1.26 + ANS(1)) $\rightarrow$ 934.138E0

The variable can also be used to do iterations, as shown in the following example. We use here the characteristic that calculators have that when you press $\boxed{\text{ENTER}}$ directly, the last command line is repeated.

Example 2.2 *We want to approximate one real root of $f(x) = x - \cos(x)$ using Newton's method to ten decimal places. The iteration principle for the method is the formula*

$$x_{n+1} = x_n - \frac{f(x_n)}{f'(x_n)} = x_n - \frac{x - \cos(x)}{1 + \sin(x)}$$

We start start setting FIX 10, Normal *modes to have the desired number of decimals. Let us start our guessing with x=-1, and introduce Newton's formula in the entry, assuming* ANS(1) *is the current value x_n:*

$\boxed{(-)}$ 1 $\boxed{\text{ENTER}}$ $\rightarrow$ -1

ans(1)-(ans(1)-cos((ans(1)))$\boxed{\div}$(1+sin(ans(1))) $\boxed{\text{ENTER}}$ $\rightarrow$ 8.7162169588

$\boxed{\text{ENTER}}$ $\rightarrow$ 2.9760606551

$\boxed{\text{ENTER}}$ $\rightarrow$ -.4257846887

$\boxed{\text{ENTER}}$ $\rightarrow$ 1.8511838176

$\boxed{\text{ENTER}}$ $\rightarrow$ .7660395196

$\boxed{\text{ENTER}}$ $\rightarrow$ .7392410675

$\boxed{\text{ENTER}}$ $\rightarrow$ .7390851386

| ENTER | → .7390851332

| ENTER | → .7390851332

Since the last value repeated itself, the sequence has converged within ten places to the approximate solution 0.7390851332.

Recalling previous answers or entries from stack

You can recall results using the up and down cursor keys to navigate through the stack, pressing ENTER once you highlight the desired item.

2.4.5 Using several previous answers or entries in calculations

Since the TI-89 has a stack for answers that are retrievable as array elements, ANS(1), ANS(2), ANS(3) and so on, it is possible to use previous answers in formulas and applications. Let us take tje following example. To enter ANS(X) for any X ≠ 1 enter 2nd   (-) , ←   ← ' *The key for* X')

Example 2.3 *Using the calculator to generate Fibbonaci's sequence,* 1, 1, 2, 3, 5, 8, 13, ..., *where n-th element is* X_n *is given by* $X_n = X_{n-1} + X_{n-2}$ *for* $n > 2$.

To do it, let us work the first eight elements in the sequence with following entries in the given order:

1 ENTER	→	1
1 ENTER	→	1
+ ANS(2) ENTER	→	2
ENTER	→	3
ENTER	→	5
ENTER	→	8
ENTER	→	13
ENTER	→	21

We will have the opportunity to use the stack of answers in the TI-89 again

2.5 Programming and User Defined Functions

The reader must read the manual to learn how to program his/her calculator, run programs and so on. Here, I limit myself to some comments and suggestions about programs and programming. Let me mention some practical points.

A) Programs need data to operate with. The TI calculators allows you to either provide the data as directly as parameters into the program or to write the code so that it asks for input while running, that is, to interact with the user. Two examples illustrating the difference are shown in section 5.2 on page 61.

2.6. A PROGRAM AS A USER DEFINED FUNCTION

Plan and program according to your preference. My personal option for circuit analysis is to avoid interactivity in the description of the circuit. That is, having the program ask for inputs that describe connections while running. My reason to do this is that unless you have set up a relative complex program, you will need to introduce again all data each time you make a mistake or you want to make a small change. Preparing all the information needed for the program before using it makes it easier for me to correct errors and make changes to the circuit.

B) Programs do not display any value unless requested to do so. Hence, when you run the program from the HOME window, you will return to the same window if and only if your program does not display or expect inputs from the keyboard. Otherwise, you are automatically taken to the IO window, where all inputs and outputs are displayed.

C) Use local variables for any variable in a program that does not need to be stored after the program stops. For example, local counters. A local variable is deleted automatically when the program stops, and therefore does not use up memory.

D) Weight the program vs. function option. Although similar in structure, differences are noticeable. On my side, if I expect more than one output, I prefer a program. If I only expect one output, I prefer to define my function.

2.6 A program as a user defined function

The TI-89 models allow the definition of functions requiring several lines using the programming environment. That is, the user has the option to define a function in a program-like way. The interested reader may look at the respective manuals.

On the other hand, it is possible to "mimic" a user defined function with a program in all cases. Simply assume the "input" variables as already defined before using the program. To illustrate, let us define a TEST1 program as shown below:

```
test1():
Prgm
:a+2 → b
:Disp B
```

Assign a value to A and then run the program. The example is trivial, but it shows the principle we want to illustrate. Other specific ways to define user functions are given next.

2.7 User defined functions

If your function is too complex, it is better to use the programming environment to define it. Refer to the guidebook on how to do this. I will explain here functions you can define on the entry line of the HOME environment.

There are two classes of functions that you can define easily directly on the command entry line: *algebraic non parametric* functions, and *algebraic parametric* functions. For the first group, if a variable Y is a function of a variable X, Y is not protected. That is, Y can be redefined with another value or expression, and the original function is lost. In the second group, the variable Y is protected and cannot be reassigned without previously deleting it. With respect to notation, non parametric functions are not distinguishable from other variables, while parametric functions are explicit in the parameters being passed. This remark will be better understood with the explanations that follow.

2.7.1 Algebraic non-parameter functions

Remember that in the TI-89 any variable that has not been assigned a numerical value, can be used as an algebraic variable. This property is used to define functions of variables in a particular way.

If you have an expression involving one or more unassigned variables (x1, x2, ...) , and you "store it" onto another variable y, [expression STO▶ y], then y becomes a function of those variables in the expression.

After the definition, each time you assign numerical values to variables x1, x2, ..., then y calculates the numerical expression obtained by substitution. This functional relation is kept as long as you do not assign values to y directly. It is important to notice that *if you assign a numerical value to y separately, the functional relation is lost*. Try the following sequences of entries on your calculator to better understand this explanation

Example 2.4 *In the first table, let us first assume that variable x is unassigned and define the "function"* $y = x^2 + 1$. *This is done on the first line. In line 2, we make x =2, and in line 3 we check that the correspondin value updated for y. Lines 4 and 5 repeat the task for x=4. You can even use lists (see next chapter), as illustrated by lines 6 and 7. In8, the value 8 is directly assigned to variable y, and the functional relationship is lost.*

Line	Entry	Answer on stack
1.	x∧2+1 STO▶ y ENTER	$x^2 + 1$
2.	2 STO▶ x ENTER	2.
3.	y ENTER	5.
4.	4 STO▶ x ENTER	4.
5.	y ENTER	17.
6.	{0, 1, 3} STO▶ x ENTER	{0. 1. 3. }
7.	y ENTER	{1. 2. 10. }
8.	8 STO▶ y ENTER	8.
9.	2 STO▶ x ENTER	2.
10.	y ENTER	8.

2.7. USER DEFINED FUNCTIONS

Notice that initially you define a one letter function y of x. Each time you change the value of x, that of y changes accordingly and you can retrieve the new value by simply entering y. Once you assign a value to y directly, the function property is lost. This can be worked with more than one variable as illustrated by the following sequences:

Line	Entry	Answer on stack
1.	a∧2+ b1-2 STO▶ y ENTER	a1^2 + b1-2
2.	4 STO▶ a1 ENTER	4.
3.	y ENTER	b1 + 14.
4.	2 STO▶ a1: 5 STO▶ b1 ENTER	5.
5.	y ENTER	17.
6.	8 STO▶ y ENTER	8.
7.	y ENTER	8.

Notice that if you assign a numeric value to one of the variables (a1 in line 2) while the other is still unassigned, update of y is done only with the assigned variable (line 3).

This feature is very handy for those situations in which you have to calculate a formula for different values which are not known beforehand, including lists or matrices,[1] but there is no need for a permanent function. If you need a permanent function, go to the following subsection.

As a matter of fact, notice that on line 4 of the second sequence we use the colon (:) to write two entries on the command line. Only the last entry goes to the stack, while other commands are also executed although not shown.

2.7.2 Functions with parameter

Defining explicitly the function with parameter, as in y(x), makes x a local variable. Now, it suffices to write something like y(2) to substitute the value of x in the function. However, now variable y is no longer available for other uses. If you want to free this variable, you must delete it. On the other side, variable x in the calculator is independent and assignments do not affect y Look at the example below. Similar comments apply to a function of several variables, like y(x,z)

Example 2.5 *The function $y(x) = x^2 + 1$ is defined with the parameter x, and the previous remarks illustrated by the following lines. The function definition uses an unassigned variable x.*

---

[1] If you know the values beforehand, you don't need a function. A list operation will do!

Line	Entry	Answer	Comment
3	x∧2+1 STO▶ y(x) ENTER	done	Define function
4	x ENTER	x	Variable x is algebraic
5	y ENTER	error message	Variable y does not exist by itself
6	y(4) ENTER	17.	For x=4 locally
7	x ENTER	x	x is algebraic
10	2 STO▶ x ENTER	2.	A value assigned to x
11	y ENTER	error message	does not affect y
12	8 STO▶ y ENTER	error message	y is not available

As previously said, he only way recover y as a variable is by deleting the function or to redefine it with a new name. To delete the function, consult your manual.

2.8 Custom Menus

When you create programs or functions, very probable you would like to write them fast on the entry line. Also, you may have a set of characters which you enter very often and it would be handy to have a means to paste them quickly on the entry line. For these situations, custom menus are the best answer. A *custom menu* is a program that allows us to paste a command or set of characters on the Entry Line using the function keys F1 F2 ... F8.

In the HOME environment, you find the toolbar at the top of the window, with submenus such as F1-Tools, F2-Algebra and so on, as shown in Fig. 2.2(a). Each environment has its own set of submenus in a tool bar. You may build and load your own custom menu. Fig. 2.2(b) shows a custom menu created to paste the functions and programs that are defined or suggested in this book. After loading a menu, pressing 2nd HOME, takes you back to the tool bar.

(a)   (b)

Figure 2.2: a) TI-89 Titanium Tool bar menu. (b)

Once you have the custom menu installed, you use the function keys to bring the name of the program or object to the command entry. For example, to enter 4 + pl(tt) you can enter 4 + F1 1 t t). The interested reader may consult the appendix in [Edward05] and the website published by [Cacovean] to create convenient custom menus to use in circuits or in other situations.

CHAPTER 3

Lists and Matrices

This chapter provides a quick review of some basics for two tools that will be very important for us: lists and matrices. As the reader might already be thinking, for our case matrices are important tools for solving sets of linear equations.

3.1 Lists

Readers who have used the calculator for Statistics may be already familiar with lists, also called arrays. In fact, the guidebooks focus on lists for Statistics and graphical purposes. But lists are indeed excellent tools for our goals, and therefore I will devote a special section to them.

A list of *dimension* n is an ordered array of n numbers or algebraic expressions between brackets ({ }), separated by commas or spaces, like

$$L = \{x_1, x_2, \ldots, x_n\} \tag{3.1}$$

A list can be defined or edited using the command line or the Data or Matrix Editor of your calculator. The reader may consult the pdf manual online for use of the Editor. You can enter the list on the command entry line directly as required. To take full advantage of lists *always* store them as variables. Individual elements in the list may be entered with operations, leaving to the calculator the required computation. It is important to remember this, since many users, particularly students, first make the calculations and then enter the results. This not only is an unnecessary inconvenience, but also a possible source of errors. However, if the individual calculations are complicated, it may be convenient to store them first in memory.

Examples of lists, their definition and storage using the command line are shown next.

List	Command Line Definition
L1 = {2, 3, 6}	{ 2 , 3 ,6 } [STO▶] L1
L2={5.18, 3.98, 6.59, -0.245}	{ 5.18 , 3.98 , 6.59 , [(-)] 0.245} [STO▶] L2
L3 = {5+3.2, 6/1.76, 3^2}	{ 5 + 3.2 , 6 ÷ 1.76 , 3 ∧ 2 } [STO▶] L3

3.1.1 Lists and your calculator

Now let us look at particular features to be considered for the use of lists in the calculator.

List variables

There are no restrictions for list variables names except those names specifically prohibited. To display a list variable on the screen enter the name on the command line.

Retrieving and displaying elements of a list:

The j-th element L_j of a list is retrieved on the command line writing the index between square brackets [ ]. For list L2 above, we have

L2[2] [ENTER] → 3.98.

Editing the list

A list can be edited on the command line by redefining the new element. For example, test the following two lines:

[(-)] 0.1 [STO▶] L 2 [2] [ENTER] → -0.1
L2 [ENTER] → {5.18 -0.1 6.59 -0.245}

Augmenting the list

We can increment the dimension n by 1 by defining a new element L_{n+1}. For example, with the previous list L1, by entering L1[4]=8 we get the new list L1 = {2,3,6,8}. If you need to add several items, you may use the editor or recall the entry that defined the list.

3.1.2 List functions and list menus

The TI-89 has a dedicated list menu which can be opened with the sequence [2nd] [5] [3]. There are several predefined list functions in the calculator. These functions can be directly typed on the entry line or else called from the menu. To select a function, enter [2nd] [5] [3] and then scroll the list or press the respective key number. Then, on the command line complete the function call typing the list (and parameters if necessary) and closing the parenthesis, if necessary, with $\boxed{)}$. It is also possible to write directly the function on the command line.

The specifics of what each of the list operations in the menu accomplishes can be consulted in the Guide Books. I will describe below only those explicitly used in the book.

seq() This command creates a list with elements in a sequence according to the parameters given in parenthesis: **seq**(*expr, var, low, high[, step]*) generates the list using the expression *expr* with respect to variable *var* for values of it between *low* and *high* with a given step. The step is optional with default 1. An example of a sequence is

seq(X²+1 ,X,2,3,0.5) $\boxed{\text{ENTER}}$ → {5.00   7.25   10.00}

sum() Yields the sum of elements in the list. For example

sum({3,2,8, 6}) → 19.

It is possible to ask for a sum in a particular range. For example, from the second to the third element,

sum({3,2,8, 6},2,3) → 10

cumsum() Returns a list of the same dimension, starting with the first element, and where consecutive elements are the sum of the previous elements in the list. For example,

cumsum({3,2,8,6}) = {3, 5, 13, 19}.

3.1.3 Working with Lists

A) Operations with a scalar: If a is a scalar, i. e., a number, and $*$ denotes an operation with numbers, then for $L = \{x_1, x_2, \ldots, x_n\}$, we have

$$a * L1 = \{a * x_1, a * x_2, \ldots, a * x_n\} \tag{3.2}$$
$$L1 * a = \{x_1 * a, x_2 * a, \ldots, x_n * a\} \tag{3.3}$$

Example 3.1 *Look at the following expressions*
3+{-3, 2, 5}   →   {0.   5.   8};
3÷{-3, 2, 5}   →   {-1.   1.5.   0.6};

{-3, 2, 5}^3   →   {-27.   8.   125};
{-3, 2, 5}/3   →   {-1.   0.667.   1.667}

B) Operations with lists: Let $L1 = \{x_1, x_2, \ldots, x_n\}$ and $L2 = \{y_1, y_2, \ldots, y_n\}$ be two lists of equal dimension, and * a binary operation for numbers. Then

$$L1 * L2 = \{x_1 * y_1, x_2 * y_2 \ldots, x_n * y_n\} \qquad (3.4)$$

whenever the individual operations are allowed.

Example 3.2 *Take* L1 = {2, -3, 4}, L2={-1.   -2.   5.}. *Then,*

```
L1+L2   →   { 1.    -5.      9.}
L1^L2   →   = {0.5   0.11111  1024.}
L1/L2   →   {-2.    1.5      0.8}
```

C) Functions: If $f(x)$ is a function and $L1 = \{x_1, x_2, \ldots, x_n\}$ is a list, then

$$f(L1) = \{f(x1), f(x2), \ldots, f(xn)\} \qquad (3.5)$$

That is, the function is applied to each element of the list. Look at the following examples.

Example 3.3 *(1) Being in radian mode, $L = \{0, \pi/4, \pi/2, -\pi/4\}$ then*

$$\sin(L) = \{0, 0.7071, 1, -0.7071\}$$

(2) Working in exact mode, let

$$f(x) = \frac{\sqrt{x}}{2+x} \quad \text{and } L = \{4, 9, -16\}$$

Then

$$f(L) = \{1/3, 3/11, -2i/7\}$$

As you might have noticed, user defined functions work fine with lists.

Example 3.4 *For a voltage divider with two resistances R1 and R2, and a source Vs, the formula for the output voltage is given by*

$$V_o = \frac{V_s \, R_2}{R_1 + R_2}$$

Let us assume that we want to find the outputs for the different combinations of values in the next table.

3.2. MATRICES

Vs (V)	R1 (Ω)	R2 (Ω)	Output (V)
5	600	2100	
8	1220	3900	
12.7	655	590	
3.1	320	780	
6	1560	4820	

Let us first define the function. Remember that "Define" in the TI-89 is obtained with `F4` `1`.

`Define vo(vs,r1,r2) = vs · r2 / (r1 + r2)` `ENTER` → `Done`
We can now define the lists:

`{5,8,12.7,3.1,6}` `STO▶` `VS`
`{600,1220,655,320,1560}` `STO▶` `R1`
`{ 2100, 3900, 590, 780, 4820 }` `STO▶` `R2`
After defining the lists, type `vo(vs,r1,r2)` `ENTER` *to get the result:*

`{3.89   6.09   6.02   2.20   4.53}`

Remark: You don't need to use the same names for the lists or variables as they are shown in the definition. You can even enter the values directly, as in `vo(5,600,2100)` to find the first result.

3.2 Matrices

Matrices play a very important role in working with linear systems. In fact, for our purposes one of the most important features that our calculator has is the ability to work with matrices and many matrix functions.

One important applications of matrices is in solving systems of linear equations. Therefore, I devote several pages to talk about this application of matrices, introducing notation and also pointing out remarks of interest for our purposes. I do not intend to enter into theoretical demonstrations, discussion of constraints and so on beyond our needs. For more detailed treatment of the theory of matrices, consult an appropriate reference.

3.3 Basic definitions and operations

Let us start with the basic definitions of interest for us in the applications. You may skip this section if you already have good knowledge of the topic.

3.3.1 Some basic definitions

A matrix **A** of order or *dimension* m by n, ($m \times n$), is a set of numbers ordered in a rectangular array of m rows and n columns. It is customary to enclose this array in parentheses, brackets, or braces. We use brackets in this book. For example, the following are matrices of order 2x2, 3x5, 2x1 and 1x3, respectively.

$$\mathbf{A} = \begin{bmatrix} 3 & 4 \\ 2 & 6 \end{bmatrix} \quad \mathbf{B} = \begin{bmatrix} 2 & -1 & 4 \\ 0 & 2 & -6 \\ 9 & 2.1 & 6 \\ 7 & 8 & -5 \\ 1 & 3 & 5 \end{bmatrix} \quad \mathbf{x} = \begin{bmatrix} 4 \\ 11 \end{bmatrix} \quad \mathbf{d} = \begin{bmatrix} 4 & 5 & 6 \end{bmatrix}$$

Matrices with only one column, like **x**, are called *column vectors*. Those with one row, like **d**, are called *row vectors*. A matrix with same number of rows and columns is a *square matrix*. **A** above is square of order 2. A *zero matrix* is one in which all elements are 0. A *unit matrix* of order n is a square matrix in which all diagonal elements are 1, and outside the diagonal are 0.

Remark: *hereafter, a column vector will be referred to simply as "vector". When necessary, distinction will be made explicitely.*

A scalar is a number, although it can also be represented by matrix of order 1x1, if you want to complicate your life. When using variables, it is customary to denote scalars by italicized or cursive, lower case letters (e.g., x), to denote vectors by bold, lower case letters as in **x**, and matrices with more than one row and one column by bold, upper case letters like **X**.

In general, to simplify notation, a matrix of order m by n can be denoted as $\mathbf{A} = [a_{ij}]_{m \times x}$. The subscript $m \times n$ may be omitted if there is no ambiguity. The notation a_{ij} represents the element in row i and column j. For vectors we may omit the number "1". In the above matrices

$$a_{12} = 2 \text{ for matrix } \mathbf{A}; \ b_{42} = 8 \text{ for matrix } \mathbf{B};$$
$$x_2 = 11 \text{ for vector } \mathbf{x}; \text{ and } d_2 = 5 \text{ for row vector } \mathbf{d}.$$

Convention *In order to be consistent with our calculator, the element in row i and column j of a matrix* **A** *will be denoted as* $\mathbf{A}[i, j]$.

3.3.2 Submatrices

A submatrix of **A** is a matrix formed by selecting from this matrix a subset of the rows and a subset of the columns of **A**, and then forming a new matrix by using those entries that appear in both the rows and columns of those selected, respecting. You can talk about deleting the rows and columns that are not in the respective submatrix. For example, the following is a submatrix a the previous **B** formed with rows 1, 3 and 5, ad columns 1 and 3. We illustrate the process by showing the deletion of the elements:

3.3. BASIC DEFINITIONS AND OPERATIONS

$$\mathbf{B} = \begin{bmatrix} 2 & \cancel{1} & 4 \\ \cancel{0} & \cancel{2} & \cancel{6} \\ 9 & \cancel{2.1} & 6 \\ \cancel{7} & \cancel{8} & \cancel{5} \\ 1 & \cancel{3} & 5 \end{bmatrix} \Rightarrow \begin{bmatrix} 2 & 4 \\ 9 & 6 \\ 1 & 5 \end{bmatrix}$$

Notice that each column (row) of a matrix $\mathbf{A}$ is a submatrix of $\mathbf{A}$.

3.3.3 Partitioning of matrices

A matrix $\mathbf{A}$ can be shown as an array of matrices, each one being a submatrix of $\mathbf{A}$. For example, the following expression shows a 5x5 matrix partitioned in submatrices $\mathbf{A}_{11}$, $\mathbf{A}_{12}$, $\mathbf{A}_{21}$, and $\mathbf{A}_{22}$.

$$\mathbf{A} = \begin{bmatrix} a_{11} & a_{12} & a_{13} & a_{14} & a_{15} \\ a_{21} & a_{22} & a_{23} & a_{24} & a_{25} \\ a_{31} & a_{32} & a_{33} & a_{34} & a_{35} \\ a_{41} & a_{42} & a_{43} & a_{44} & a_{45} \\ a_{51} & a_{52} & a_{53} & a_{54} & a_{55} \end{bmatrix} = \begin{bmatrix} \mathbf{A}_{11} & \mathbf{A}_{12} \\ \mathbf{A}_{21} & \mathbf{A}_{22} \end{bmatrix} \quad (3.6)$$

where

$$\mathbf{A}_{11} = \begin{bmatrix} a_{11} & a_{12} & a_{13} \\ a_{21} & a_{22} & a_{23} \\ a_{31} & a_{32} & a_{33} \end{bmatrix} ; \mathbf{A}_{12} = \begin{bmatrix} a_{14} & a_{15} \\ a_{24} & a_{25} \\ a_{34} & a_{35} \end{bmatrix} ; \mathbf{A}_{21} = \begin{bmatrix} a_{41} & a_{42} & a_{43} \\ a_{51} & a_{52} & a_{53} \end{bmatrix}$$

and

$$\mathbf{A}_{22} = \begin{bmatrix} a_{44} & a_{45} \\ a_{54} & a_{55} \end{bmatrix}$$

Particular cases of partitions are those with the column vectors $\mathbf{a}_j$ and row vectors $\mathbf{a}^{(j)}$. If $\mathbf{A}$ is of order $m \times n$,

$$\mathbf{A} = \begin{bmatrix} \mathbf{a}_1 & \mathbf{a}_2 & \cdots & \mathbf{a}_n \end{bmatrix} \text{ and } \mathbf{A} = \begin{bmatrix} \mathbf{a}^{(1)} \\ \mathbf{a}^{(2)} \\ \vdots \\ \mathbf{a}^{(m)} \end{bmatrix} \quad (3.7)$$

We will use these forms for explanations later.

3.3.4 One remark on multiplication of matrices

I leave to the reader to review the matrix operations of addition, subtraction, multiplication, scalar multiplication, and transposition. I only want to bring to your attention a property of multiplication. Namely, when you multiply two matrices, we have the following property:

$$\mathbf{A}\,\mathbf{B} = [\mathbf{A}\,\mathbf{b}_1 \quad \mathbf{A}\,\mathbf{b}_2 \cdots \mathbf{A}\mathbf{b}_n] \tag{3.8}$$

That is, if $\mathbf{b}_j$ is the j-th column vector of $\mathbf{B}$, then $\mathbf{A}\mathbf{b}_j$ is the j-th column vector of the product $\mathbf{A}\cdot\mathbf{B}$.

3.3.5 Matrices and linear combinations

A linear combination of variables $z_1, z_2, \ldots, z_n$ is an expression of the form

$$a_1\,z_1 + a_2\,z_2 + \ldots a_n\,z_n \tag{3.9}$$

which can be expressed as a multiplication of a row vector for the coefficient a_i and a column vector for the variables z_i as

$$[a_1 \quad a_2 \cdots a_n] \begin{bmatrix} z_1 \\ z_2 \\ \vdots \\ z_m \end{bmatrix} \tag{3.10}$$

Now, for a set of linear combinations we can express the set either as a linear combination with vector coefficients or as a matrix by a vector multiplication. This duality allows us to develop the theory to use practical procedures to facilitate our circuit calculations. The development is shown next in expression (3.11) using three variables for easy reading, but can be expanded to any number of variables without problem

$$\begin{bmatrix} a_{11}\,z_1 + a_{12}\,z_2 + a_{13}\,z_3 \\ a_{21}\,z_1 + a_{22}\,z_2 + a_{23}\,z_3 \\ a_{31}\,z_1 + a_{32}\,z_2 + a_{33}\,z_3 \end{bmatrix} = \begin{bmatrix} a_{11} \\ a_{21} \\ a_{31} \end{bmatrix} z_1 + \begin{bmatrix} a_{12} \\ a_{22} \\ a_{32} \end{bmatrix} z_2 + \begin{bmatrix} a_{13} \\ a_{23} \\ a_{33} \end{bmatrix} z_3 \tag{3.11a}$$

and

$$\begin{bmatrix} a_{11}\,z_1 + a_{12}\,z_2 + a_{13}\,z_3 \\ a_{21}\,z_1 + a_{22}\,z_2 + a_{23}\,z_3 \\ a_{31}\,z_1 + a_{32}\,z_2 + a_{33}\,z_3 \end{bmatrix} = \begin{bmatrix} a_{11} & a_{12} & a_{13} \\ a_{21} & a_{22} & a_{23} \\ a_{31} & a_{32} & a_{33} \end{bmatrix} \begin{bmatrix} z_1 \\ z_2 \\ z_3 \end{bmatrix} \tag{3.11b}$$

Now, let us bring this to a practical interpretation which in fact will become a powerful tool. The vector of linear combinations in the left is expressed in the form

$$\mathbf{a}_1 z_1 + \mathbf{a}_2 z_2 + \mathbf{a}_3 z_3 = [\mathbf{a}_1 \quad \mathbf{a}_2 \quad \mathbf{a}_3]\mathbf{z} = \mathbf{A}\mathbf{z}$$

where $\mathbf{a}_j$ is the column vector for the coefficients of z_j in the set of linear combinations, and $\mathbf{A}$ is the matrix formed with those columns. Now, if we premultiply by a matrix $\mathbf{B}$, we have

$$\mathbf{B} \begin{bmatrix} a_{11} z_1 + a_{12} z_2 + a_{13} z_3 \\ a_{21} z_1 + a_{22} z_2 + a_{23} z_3 \\ a_{31} z_1 + a_{32} z_2 + a_{33} z_3 \end{bmatrix} = \mathbf{B}\,\mathbf{a}_1 z_1 + \mathbf{B}\,\mathbf{a}_2 z_2 + \mathbf{B}\,\mathbf{a}_3 z_3 \qquad (3.12)$$
$$= [\mathbf{B}\,\mathbf{a}_1 \quad \mathbf{B}\,\mathbf{a}_2 \quad \mathbf{B}\,\mathbf{a}_3]\,\mathbf{z} = \mathbf{BAz}$$

In other words, don't worry too much about the presence of variables. The columns of **B A** will tell you the coefficients for each variable in the set of linear combinations that results after multiplication. We will see that this is a very valuable property for us!

3.4 Matrices and Calculators

3.4.1 Matrix menu

For an efficient work with the calculator, you will very often need to open the matrix menu. Enter [2nd] [5] [4] and then scroll the list or press the respective key number. Some of the functions in the menu will be explained later.

3.4.2 Matrix variables and operations

There are no restrictions for matrix variables names in the TI-89, except those names specifically prohibited in the manual. The typical matrix and scalar operations defined in textbooks may be entered directly on the entry line, provided dimensions are compatible. These operations are :

Addition and subtraction: A + B, A - B

Matrix multiplication: A · B

Multiplication or division by a scalar m: $m \cdot \mathbf{A}$, $\mathbf{A}/m$

Raise to a power for a square matrix: $\mathbf{A}^m$

Inverse of a square matrix $\mathbf{A}^{-1}$

Functions and transformations are explained later.

Dot operations in the TI-89

In the TI-89 dot operations, or element operations, can be executed. A dot followed by an operation key, (.<OP>), between two matrices $\mathbf{A} = [a_{ij}]$ and $\mathbf{B} = [b_{ij}]$ of similar dimensions, yields a new matrix $\mathbf{C} = [c_{ij}]$ where $c_{ij} = a_{ij}\,OP\,b_{ij}$. For example, if

then
$$\mathbf{A} = \begin{bmatrix} 3 & 4 \\ 2 & 6 \end{bmatrix} \qquad \mathbf{B} = \begin{bmatrix} 2 & -1 \\ 0 & 2 \end{bmatrix}$$

$$\mathbf{A}.*\mathbf{B} = \begin{bmatrix} 6 & -4 \\ 0 & 1 \end{bmatrix}; \qquad \mathbf{A}.\wedge\mathbf{B} = \begin{bmatrix} 4 & -0.25 \\ 1 & 36 \end{bmatrix}$$

Dot operations are also valid between a matrix and a scalar:

$$\mathbf{A}.+5 = \begin{bmatrix} 8 & 9 \\ 7 & 6 \end{bmatrix}; \qquad \mathbf{A}.\wedge 2 = \begin{bmatrix} 9 & 16 \\ 4 & 36 \end{bmatrix}$$

3.4.3 Creating matrices

Matrices in the calculator can be created using the editor tool provided by the calculator or on the command line. Whatever method you choose, don't worry too much about matrix elements with operations involved. Just write them while entering the elements, and let the calculator do the work. If the expression is too complicated, I suggest to first work the expression and store it to a variable which you will later enter where needed.

Using the calculator editor tool

Press APPS and open Data/Matrix Editor. After defining your matrix, press HOME or 2nd ESC to go back to home or to the app where you were before invoking the editor.

Using the calculator editor is almost straight forward. Consult you guidebook.

On the command line

A matrix defined on the command line must be enclosed in brackets. You can either enter each row in brackets with elements separated by commas, as in

[[1,2][3,4]] ENTER .

or use commas to identify elements in a row and semicolon (;) to separate rows. The same matrix can be thus entered with

[1,2 ; 3,4] ENTER .

Let us now create the more complicated matrix below. I have written coefficients with explicit operations on purpose, to stress that we can enter the operations directly and let the calculator do them. Let us now solve with the calculator.

$$\begin{bmatrix} (35 + 126.7) & -\frac{72}{17} & 127 \\ \frac{189}{3.78} - 2.13 \times 4 & 0 & 115 \\ 3 & 44 & -11 \end{bmatrix} \qquad (3.13)$$

3.4. MATRICES AND CALCULATORS

Separating the elements in the row with a comma (,) and the rows with semi-colon (;), this matrix **A** can be entered as:

[35+126.7, (-) 72/17,127; 189/3.78-2.13 × 4, 0, 115; 3,44, (-) 11] STO▶ a

With row vector also between brackets, the matrix would be entered as:

[[35+126.7, (-) 72/17,127] [189/3.78-2.13 × 4,0,115] [3,44, (-) 11]] STO▶ a

The result in any case, using FLOAT 5 as setting, is

$$\begin{bmatrix} 161.7 & -4.2353 & 127 \\ 41.48 & 0 & 115 \\ 3 & 44 & -11 \end{bmatrix} \qquad (3.14)$$

Notice that the commas that you enter to separate elements in rows are not displayed on output.

3.4.4 Retrieving and editing elements

To retrieve an the element at row i and column j of matrix a enclose the indices between brackets. Using (3.14),

a[k,j]:   a[3,2]   → 44

You can edit an element in a matrix directly on the entry line by storing the new value to the element. Try the following:

5 STO▶ a[1,2]:a ENTER

3.4.5 Retrieving and editing rows and columns

The row vector k of matrix a is denoted as a[k]. Using (3.14) for example,

a[2]   →   [41.48   0   115]

Using the transpose function from the matrix menu it is possible to retrieve columns. However, the result may be a row vector or a column vector, depending on how you enter the command. Again, using the same matrix we already have see the difference between $a^T[2]$ and $a^T[2]^T$. The transpose operation is item 1 in the TI-89 matrix menu.

You can also retrieve a column vector with the use of submatrices, as explained in the next section.

To edit a complete row, simply store the new row using the same notation, as in

[-3   1   105] STO▶ a[2]:   a ENTER will yield, assuming that **A** is the original in (3.14),

$$\begin{bmatrix} 161.7 & -4.2353 & 127 \\ -3 & 1 & 105 \\ 3 & 44 & -11 \end{bmatrix}$$

If you want to edit a column, first transpose the matrix so you can work with the rows. For example, starting with the same matrix **A** from (3.14),

a^T STO▶ a: [-3 1 105] STO▶ a[2]: a^T STO▶ a: a ENTER will yield

$$\begin{bmatrix} 161.7 & -3 & 127 \\ 41.48 & 1 & 115 \\ 3 & 105 & -11 \end{bmatrix}$$

3.4.6 Submatrices

A submatrix can be obtained by deleting appropriate rows and columns. In our calculator, we can do this either using the Matrix Editor or the submatrix function subMat().

Deleting rows and columns

To delete a row or a column of a matrix using the matrix editor, place the cursor on any element of the row or column to delete and then press 2nd F1 2 2 to delete the row, or 2nd F1 2 3 to delete the column.

The SubMat function

The subMat() function in matrix menu allows us to define submatrices of continuous rows and columns. This function is option G in the menu, and can be called with 2nd 5 4 ALPHA 7.

The format for the function is subMat(matrix, j,k,l,m). Here, j,k are the row and column of the first element in the submatrix, and l,m that of the last element. If only two parameters are provided, those are of the first element, the last element being the one at the last column and row default. Using again (3.14) for our reference

subMat(a, 1,2,2,3) and subMat(a, 1,2,2,2)

yield, respectively,

$$\begin{bmatrix} -4.2353 & 127 \\ 0 & 115 \end{bmatrix} \text{ and } \begin{bmatrix} -4.2353 \\ 0 \\ 44 \end{bmatrix}$$

Observe that to retrieve the vector column j from matrix $\mathbf{A}_{m \times n}$ we write subMat(a,1,j,n,j).

3.4. MATRICES AND CALCULATORS

3.4.7 Matrix functions in the matrix menu

When you display the matrix menu in your calculator, several functions are available. Those of interest for us are described next. The reader can consult the guidebook for a complete description of the other functions. Also, many list functions are applicable to matrices.

- 1. *Transposed* function `1.` `T`, to obtain the transposed of a matrix

- 2. *Determinant* function `det(`, to obtain the determinant of a square matrix. In this book, this function is used mainly to check the validity of a solution process in programming. It has of course other uses.

- 4. *Reduced row-echelon form* function `rref(`, of interest to us to solve systems of linear equations, as illustrated in next section.

- 7. *Augment function* `augment(`. This function can be used to create matrices from blocks. See subsection 3.4.7 on the following page and the example there.

- F. *New Matrix* function `newMat(`. Typing `newMat(a,b)` on the entry line produces a zero matrix of order $a \times b$. In this book, it is used mainly in initiation steps for programming.

- G. *Submatrix* function `submat(`, already illustrated in the previous section.

- I. *Dimensions.* This menu item has three functions: `dim(`, `rowDim(`, and `colDim(`. The first one yields a list with the number of rows and number of columns for the matrix. They are of interest to us mainly for programming purposes.

- J. *Row operations*: These are the following. Notice that the result is not stored to a variable other than ANS. Thus, if you need to work with the new matrix, store it onto a variable.

    1. `rowSwap(a,j,k)` yields where rows j and k in matrix a have been swapped.
    2. `rowAdd(a,j,k)` adds row j to row k for matrix a. It is equivalent to `a[j]+a[k]` STO▶ `a[k]`.
    3. `mRow(m,a,j)` multiplies row j of matrix a by m. It is equivalent to `m*a[j]` STO▶ `a[k]`.
    4. `mRowAdd(m,a,j,k)` multiplies row j of matrix a by m and adds the result to row k. It is equivalent to `m*a[j]+a[k]` STO▶ `a[k]`.

Column operations There are no column operations included in the menu. These can be done by first transposing the matrix, apply the row operations, and transpose back the result. For example, to swap columns 1 and 2, we could enter a^T STO▶ a: rowSwap(a,1,2) STO▶ a: a^T STO▶ a.
We could also define column functions as follows.

1. Swapping columns: Define colSwap(a,j,k) = (rowSwap(a^T,j,k))^T

2. Adding columns: Define colAdd(a,j,k) = (rowAdd(a^T,j,k))^T

3. Multiplying a column by a constant: Define mCol(m,a,k) = (mRow(m, a,^{T}k))^T

4. Adding m times column j to column k: Define mColAdd(m,a,j,k) = (mRowAdd(m,a^T,j,k))^T

Using augment(function

The augment function is useful in many ways. Among other uses, it allows us to build a matrix using other matrices as blocks. That is, given matrices **A**, **B**, **C**, and **D**, of appropriate dimensions, it is possible to construct matrices where these ones are submatrices, of the forms

$$[\mathbf{A} \ \mathbf{B}] \quad \begin{bmatrix} \mathbf{A} \\ \mathbf{C} \end{bmatrix} \quad \begin{bmatrix} \mathbf{A} & \mathbf{B} \\ \mathbf{C} & \mathbf{D} \end{bmatrix}$$

The command augment(a,b) will produce the first form only, that is, [**A** **B**]. With a semicolon, augment(a;c) produces the second form.

More complex forms, like the third one shown above, can be built by steps, as illustrated with the following example.

Example 3.5 *Assume we have already the following matrices, stored in* a, b, c, d

$$\mathbf{A} = \begin{bmatrix} 1 & 2 \\ 3 & 4 \end{bmatrix}; \quad \mathbf{B} = \begin{bmatrix} -1 & -2 \\ -3 & -4 \end{bmatrix}; \quad \mathbf{C} = \begin{bmatrix} 5 & 6 \end{bmatrix}; \quad \mathbf{D} = \begin{bmatrix} 9 & 10 \end{bmatrix}$$

Then

augment(a,b) ENTER $\to \begin{bmatrix} 1 & 2 & -1 & -2 \\ 3 & 4 & -3 & -4 \end{bmatrix}$

augment(a;c) ENTER $\to \begin{bmatrix} 1 & 2 \\ 3 & 4 \\ 5 & 6 \end{bmatrix}$

Let us use the colon (:) to enter several commands with one line next. You can build now the matrix:

augment(a;c) `STO▶` f: augment(b;d) `STO▶` g:

augment(f,g) `ENTER` $\rightarrow \begin{bmatrix} 1 & 2 & -1 & -2 \\ 3 & 4 & -3 & -4 \\ 5 & 6 & 9 & 10 \end{bmatrix}$

3.5 Matrices and Linear equations

Matrices and systems of linear equations are intimately associated since ancient times. To simplify discussion and space, I work here with 3 variables. When the generalization to any number of variables is not straightforward, or particular remarks must be done for more or less variables, I will discuss the particular case. For now, the objective is to introduce notation and talk about dealing with matrices with the calculator.

$$\begin{aligned} a_{11}x_1 + a_{12}x_2 + a_{13}x_3 &= b_1 \\ a_{21}x_1 + a_{22}x_2 + a_{23}x_3 &= b_2 \\ a_{31}x_1 + a_{32}x_2 + a_{33}x_3 &= b_3 \end{aligned} \qquad (3.15)$$

which can be expressed in matrix equation as

$$\begin{bmatrix} a_{11} & a_{12} & a_{13} \\ a_{21} & a_{22} & a_{23} \\ a_{31} & a_{32} & a_{33} \end{bmatrix} \begin{bmatrix} x_1 \\ x_2 \\ x_3 \end{bmatrix} = \begin{bmatrix} b_1 \\ b_2 \\ b_3 \end{bmatrix} \qquad (3.16)$$

The 3x3 matrix is the coefficient matrix **A**, formed with the coefficients of the variables. Observe that the rows in this matrix are the coefficients in the equations, while the columns are the coefficients of variables in different equations. Using labels, we illustrate this as follows

$$\mathbf{A} = \begin{matrix} & \begin{matrix} x_1 & x_2 & x_3 \end{matrix} \\ \begin{matrix} \text{Eq. 1} \\ \text{Eq. 2} \\ \text{Eq. 3} \end{matrix} & \begin{bmatrix} a_{11} & a_{12} & a_{13} \\ a_{21} & a_{22} & a_{23} \\ a_{31} & a_{32} & a_{33} \end{bmatrix} \end{matrix} \qquad (3.17)$$

I will use labels for rows and columns to explain the structure of the matrices, simplify explanations and so on.

On the right hand side of (3.16) we find the vector of knowns. Instead of a vector, it could be a matrix, as illustrated in subsection 3.5.2. This vector or matrix of knowns can be generated for different reasons and depending on the application.

3.5.1 Solving linear equations

To solve system (3.16), we may work in two forms with the calculator using matrices:

(a) Generate the matrix **A** and the vector **b** and enter on the command line the sequence

A ∧ (-) 1 × b

The vector that results is the desired **x**.

(b) Generate the augmented matrix **C** = [**A** **b**] and apply the reduced row echelon form transformation (**rref(C)**), reading the solution from the last column. Both methods are illustrated next.

Example 3.6 *Solve the following set of equations:*

$$
\begin{aligned}
(35 + 126.7)\, x_1 - \tfrac{72}{17} x_2 + 127\, x_3 &= 109 \\
\left(\tfrac{189}{3.78} - 2.13 \times 4\right) x_1 + 0\, x_2 + 115\, x_3 &= -87 \\
3\, x_1 + 44\, x_2 - 11\, x_3 &= -7
\end{aligned}
$$

The coefficient matrix is the same as in expression (3.13) on page 30. The vector of knowns **B** *may be similarly created. Once back at the home environment if you used the editor, the sequence*

A ∧ (-) 1 × B ENTER

will produce the display

$$
\begin{bmatrix} 1.74673 \\ -.634827 \\ -1.38656 \end{bmatrix}
$$

which is interpreted as the desired result: $x_1 =$ 1.74673, $x_2 =$ -634827, *and* $x_3 =$ -1.38656.

On the other hand, using **rref(** *together with the* **augment(** *transformation:*

rref(augment(A,B)) ENTER

yields

$$
\begin{bmatrix} 1 & 0 & 0 & 1.74673 \\ 0 & 1 & 0 & -.634827 \\ 0 & 0 & 1 & -1.38656 \end{bmatrix}
$$

The last column is the solution.

3.5. MATRICES AND LINEAR EQUATIONS

Personally, I prefer to use the first method just because the unit submatrix make it harder to read the result. However, sometimes it is easier to deal with the row reduced echelon form procedure.

3.5.2 Multiple systems and knowns. Matrix B instead of vector b

There are several situations in which a matrix **B** is used in the above process. The interpretation of the results depend on the application. I illustrate with two examples.

Example 3.7 *Take two systems in which the only difference is the knowns-vector:*

$$\begin{array}{ll} 3x + 2y = 8 & \qquad 3x + 2y = 1 \\ -x + 3y = 1 & \qquad -x + 3y = 7 \end{array}$$

Now create the coefficient matrix **A** *for the equations, and the matrix* **B** *using the vectors as columns:*

$$\mathbf{A} = \begin{bmatrix} 3 & 2 \\ -1 & 3 \end{bmatrix}$$

and

$$\mathbf{B} = \begin{bmatrix} 8 & 1 \\ 1 & 7 \end{bmatrix}$$

Then

$$\mathbf{A}^{-1}\mathbf{B} = \begin{bmatrix} 3 & 2 \\ -1 & 3 \end{bmatrix}^{-1} \times \begin{bmatrix} 8 & 1 \\ 1 & 7 \end{bmatrix} = \begin{bmatrix} 2 & -1 \\ 1 & 2 \end{bmatrix} \qquad (3.18)$$

The result means that $x = 2$ *for the first set, and -1 for the second one, while the respective solutions for y are 1 and 2.*

On the other hand, the entry `rref(augment(A B))` $\boxed{\text{ENTER}}$ *yields*

$$\begin{bmatrix} 1 & 0 & 2 & -1 \\ 0 & 1 & 1 & 2 \end{bmatrix}$$

and we now read the solutions from the submatrix at the right of the unit submatrix of the left.

Let us look at another situation with the same matrices:

Example 3.8 *Take the system*

$$\begin{array}{l} 3x + 2y = 8 + z \\ -x + 3y = 1 + 7z \end{array}$$

We see that the right hand side of the system is a set of linear combinations that can be expressed as $\mathbf{B}\begin{bmatrix} 1 & z \end{bmatrix}^T$, *where* **B** *is the same matrix of the previous example.*

Using the property illustrated by expression (3.12) on page 29, we can proceed to multiply $\mathbf{A}^{-1}\mathbf{B}$, which yielded the result (3.18).

The important step here is interpretation. Although the matrix result now is the same as in the previous example, the situation is different and now we see the solution for each variable from the rows: The first row means x = 2 - z, and the second row means y = 1 + 2z.

Notice that it is not only a matter of entering data onto the calculator. You must interpret your results!

3.5.3 Exchanging knowns and unknowns

In subsection 3.3.4 I mentioned the fact that the multiplication of a matrix $\mathbf{A}$ and a column vector $\mathbf{x}$ can be expressed as a linear combination

$$\mathbf{a}_{(1)} x_1 + \mathbf{a}_{(2)} x_2 + \ldots \mathbf{a}_{(n)} x_n$$

where $\mathbf{a}_{(m)}$ is the m-th column of the coefficient matrix $\mathbf{A}$ and $x_1, x_2, \ldots$ are the elements of vector $\mathbf{x}$. Let us now use this property to illustrate a very useful procedure that allows us to exchange "knowns and unkowns" in a problem expressed as a set of equations as illustrated next.

The system of equations in example 3.8 can also be expressed as

$$\mathbf{a}_{(1)} x + \mathbf{a}_{(2)} y = \mathbf{b}(1) + \mathbf{b}(2) z \qquad (3.19)$$

In this system it is assumed that variable z is known. In practical terms, we say that we want to express x and y as functions of z. Now, imagine that there is a situation in another problem where where x is known and z unknown, and yet the relational equations are the same as before and you already have the matrices $\mathbf{A}$ and $\mathbf{B}$ available. How can we proceed?

One way is to rewrite the new set of equations. This is equivalent to exchange, with proper sign modifications, the respective columns from both matrices so we can represent the new problem as

$$-\mathbf{b}_{(1)} z + \mathbf{a}_{(2)} y = \mathbf{b}_{(1)} - \mathbf{a}_{(2)} x$$

Let us illustrate the whole process with the numerical example given before.

Example 3.9 *The original set is*

$$\begin{aligned} 3x + 2y &= 8 + z \\ -x + 3y &= 1 + 7z \end{aligned}$$

and what we want to solve now is

$$\begin{aligned} -z + 2y &= 8 - 3x \\ -7z + 3y &= 1 + x \end{aligned}$$

We already have the matrices $\mathbf{A}$ and $\mathbf{B}$ from the original set:

3.5. MATRICES AND LINEAR EQUATIONS

$$\mathbf{a} = \begin{bmatrix} 3 & 2 \\ -1 & 3 \end{bmatrix} \text{ and } \mathbf{b} = \begin{bmatrix} 8 & 1 \\ 1 & 7 \end{bmatrix}$$

Our objective is to manipulate these matrices so we can solve the new set. I show the steps below, omitting the ENTER *key at the end of each command.*

Remember that the calculator does not work with or access columns. Since it can work with rows, we must first transpose our matrices and work with the corresponding rows. Then we transpose again to get the desired result. Let us first use the embedded functions in our TI-89. After that, let us make the changes ourselves by retrieving and editing the rows.

Step 1: Create the transposed of the augmented matrix C=[A, B]
^{T} (augment(a,b))^T STO▶ c

Step 2: Exchange rows 1 and 4 of C *This amounts to exchanging columns 1 and 4 from the original augmented matrix* **C**:
rowSwap(c,1,4) STO▶ c

Step 3: Multiply row 1 by -1, and repeat with row 4 : mrow((-) 1,c,1)
STO▶ c:mrow((-) 1,c,1) STO▶ c

Step 4: Transpose [C *again]: This will yield the desired set of equations in augment form:* c^T STO▶ c

Step 5: Apply RREF function to solve rref(c) ENTER

The last step results in the matrix

$$\begin{bmatrix} 1 & 0 & 2 & -1 \\ 0 & 1 & 5 & -2 \end{bmatrix}$$

Hence, $z = 2 - x$ *and* $y = 5 - 2x$.

Let us now work it by retrieving and editing the rows, using an auxiliary variable or the stack.

Step 1: transpose: a^T STO▶ a : b^T STO▶ b

Step 2: retrieve "column" 1 of A and store: a[1] STO▶ x

Step 3: Exchange "columns" with sign changes:
(-) b[2] STO▶ a[1] : (-) x STO▶ b[2]

Step 4: transpose again: a^T STO▶ a : b^T STO▶ b

Step 5: solve a^{-1} * b

The example chosen was simple enough for the reader to check by hand calculations. This procedure, however, becomes very handy in many situations, as we will see in later chapters.

CHAPTER 4

Four Network Theorems and Applications

You may skip this chapter and come back later if you need refreshing. It is of theoretical nature, but very practical consequences. The theorems presented in this chapter provide the theoretical background for many applications of our calculator. Remark that the applications themselves are doable with pencil and paper, but that's another story.

There are four very important theorems for circuits: a) the *Voltage Source substitution theorem*, the *Current Source Substitution Theorem*, applicable to all circuits, and b) the *Homogeneity Theorem* and the *Superposition Theorem* for linear circuits. In this chapter I also use the first three theorems to illustrate how we can combine theory with calculator, and obtain results for situations in the same circuit with different sources or information. A full chapter is devoted later to superposition theorem.

4.1 Substitution Theorems

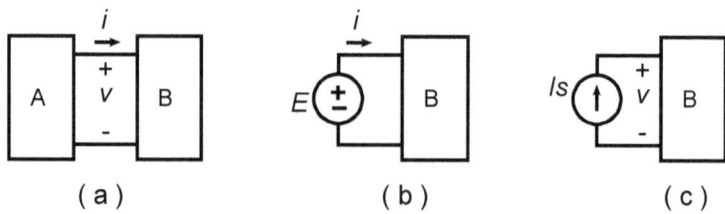

Figure 4.1: Substitution Theorems: (a) Case; (b) Voltage Substitution; (c) Current Substitution

The substitution theorems state the following:

4.1. SUBSTITUTION THEOREMS

Voltage Source Substitution Theorem: Let a circuit be partitioned in two subcircuits A and B as shown in Fig. 4.1(a). If the voltage v is known, say $v = E$, than subcircuit A may be substituted by a voltage source of value E without any change of currents and voltages in subcircuit B, including the current i entering the subcircuit. This is illustrated in Fig. 4.1(b)

Current Source Substitution Theorem: Let a circuit be partitioned in two subcircuits A and B as shown in Fig. 4.1(a). If the current i is known, say $i = Is$, than subcircuit A may be substituted by a voltage source of value Is without any change of currents and voltages in subcircuit B, including the voltage v at the terminals of the subcircuit. This is illustrated in Fig. 4.1(c)

Let us start with an example

Example 4.1 *An analysis of the circuit in Fig. 4.2(a) yields the following results:*

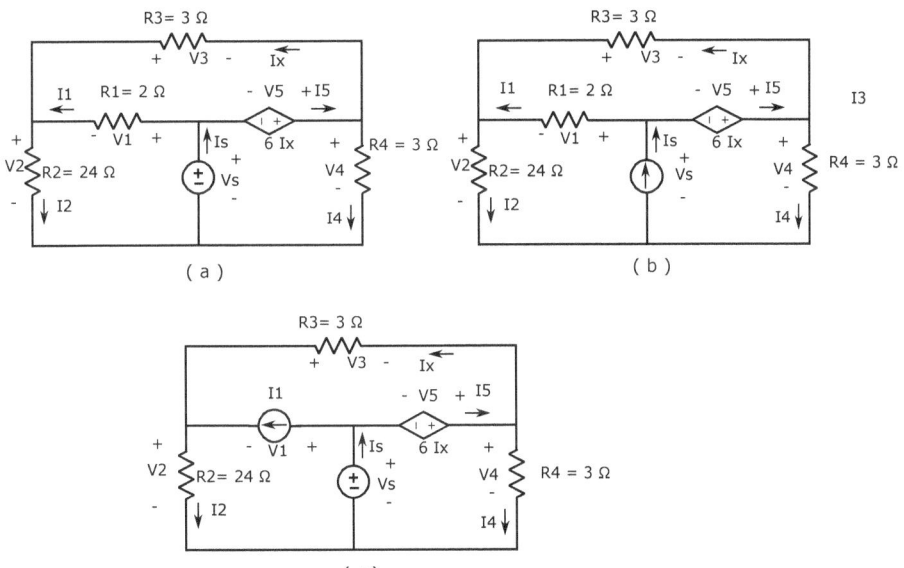

Figure 4.2: Example of Substitution Theorems: (a) Original circuit; (b) Voltage source substituted by current source; (c) Current Substitution of resistor R1

$$\begin{array}{lll}
Vs = 30 \text{ V} & Is = 7 \text{ A} & V1 = 6 \text{ V} \quad I1 = 3 \text{ A} \\
V2 = 24 \text{ V} & I2 = 1 \text{ A} & V3 = 6 \text{ V} \quad I3 = 2 \text{ A } (= -I\phi) \\
V4 = 18 \text{ V} & I4 = 6 \text{ A} & (6I\phi) \ V5 = -12 \text{ V} \quad I5 = 4 \text{ A}
\end{array}$$

In inset (b), the voltage source has been substituted by a current source. A current source is also used in Fig. 4.2(c) substituting a resistance. The reader can verify that the voltages and currents are equal in all cases.

42   CHAPTER 4. FOUR NETWORK THEOREMS AND APPLICATIONS

Remark: As it can be appreciated in the previous example, when going from circuit (a) to circuit (b), it is actually not important from the numerical point of view wheter you use a voltage or a current source.

A more interesting application of the substitution theorem can be seen in the next example.

Example 4.2 *Consider the configuration in Fig. 4.3(a), where block A is a resistive linear subnetwork and the black box element is of any type, like for a example a capacitance or a non linear element.*

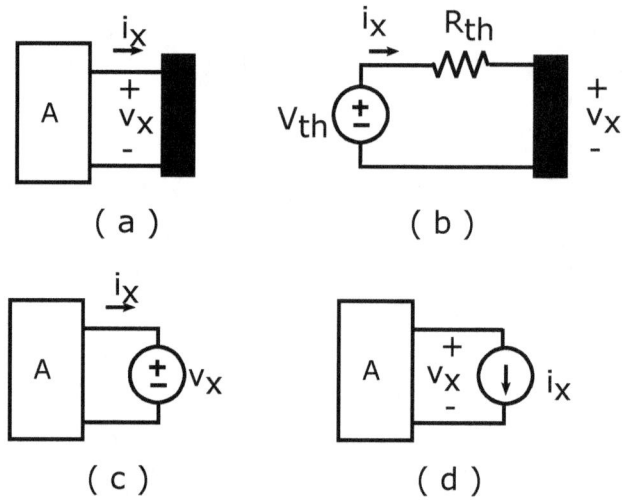

Figure 4.3: Another Example of Substitution Theorems: (a) Original circuit; (b) Block A substituted by its Thevenin's equivalent; (c) and (d) Substitution theorems applied

Fig. 4.3(b) shows the substitution of block A by its Thevenin's equivalent. This configuration is much easier to dealt with, since it reduces to two equations:

$$\text{Loop equation:} \quad V_{th} = R_{th} i_x + v_x$$

and

$$\text{Element's equation:} \quad f(v_x, i_x) = 0$$

The latter may be non linear, or may involve differentials, as it is the case of capacitances and inductances. Whatever the case, it is possible to obtain a solution by some means, including the use of calculator, and find the values for v_x and i_x.

Once these values are found, the black box may be substituted by either a voltage or a control source in the original circuit, as shown in Fig. 4.3(c) and (d), respectively. From this circuit we may solve for the elements in block A. This is a very common way to proceed, and I will show later an example.

4.2 Homogeneity and Proportionality

Consider a circuit with only one independent source, as illustrated by Fig. 4.4. The source may be of any type. If it is a voltage source, then v_s is known and i_s is to be calculated, and vice versa. In fact, except for situations where a mathematical inconsistency may arise, the treatment of these magnitudes is a matter of mathematical convenience, as pointed out in previous remark. For general discussion, we shall speak of a source z, without any particular reference to the type of magnitude, except if confusion may arise.

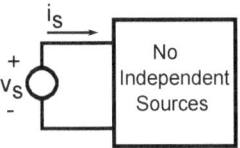

Figure 4.4: One Source Circuit

The circuits we are dealing with are linear. Because of this, the next theorem applies to them.

Theorem of Homogeneity: *In a linear network with only one independent source of value z, (Fig. 4.4), when this source is substituted by one of value Kz, where K may be a constant, every then every current and voltage in the circuit will be multiplied by K.*

Mathematically, this is usually stated in linear systems, considering y an output and z an input, as follows:

$$\text{If the system is linear an } y = f(z), \text{ then } y_1 = f(Kz) = k\,f(z) = ky \quad (4.1)$$

REMARK: If the circuit in Fig. 4.4 contains only linear resistors, linear dependent sources, and operational amplifiers, then the principle of homogeneity is also valid when K is a time function. This special case is not applicable to circuits containing elements that are either time dependent or whose characteristic may involve time dependence, as it happens with capacitances or inductances.

For linear resistive circuits, we also have the Principle of Proportionality, from which that of Homogeneity is a consequence. This principle can be stated as follows:

Principle of Proportionality: *If a linear resistive network has only one independent source z, then every current and voltage in the circuit is of the form Kz, where K is a constant which depends exclusively on the circuit elements, but not on the source.*

Let us look at an example.

Example 4.3 *Consider the circuit from example 4.1, namely, that of Fig. 4.3(a). If the source V_s is now $24e^{-100t}$ V, then every current and voltage should be multiplied by $0.8e^{-100t}$, which is the factor needed to convert the original 30 V source to the new one. The result is then*

$$
\begin{array}{ll}
Vs = 24e^{-100t} \text{ V} & Is = 5.6e^{-100t} \text{ A} \\
V1 = 4.8e^{-100t} \text{ V} & I1 = 2.4e^{-100t} \text{ A} \\
V2 = 9.6e^{-100t} \text{ V} & I2 = 0.8e^{-100t} \text{ A} \\
V3 = 4.8e^{-100t} \text{ V} & I3 = 1.6e^{-100t} \text{ A } (= -I\phi) \\
V4 = 13.6e^{-100t} \text{ V} & I4 = 4.8e^{-100t} \text{ A} \\
(6I\phi)\ V5 = -9.6e^{-100t} \text{ V} & I5 = 3.2e^{-100t} \text{ A}
\end{array}
$$

Furthermore, if we want a specific voltage, say V_4 to be of another value A, then we multiply all elements, including the source, by $A/6$, so we have the required source value as well as the new currents and voltage.

We can, in general take a matrix or a list to have an easier way to calculate. Take for example

$$
\mathbf{M} = \begin{bmatrix} 30 & 7 & 6 & 3 \\ 12 & 1 & 6 & 2 \\ 18 & 6 & -12 & 4 \end{bmatrix}
$$

to represent the values of the original circuit. Changes are done in one step with the operation $K\mathbf{M}$

4.3 Superposition Theorem

This theorem is valid for all linear systems, where it is call additivity property. In our circuits, it is applied when several independent sources are present, as illustrated by Figure 4.5.

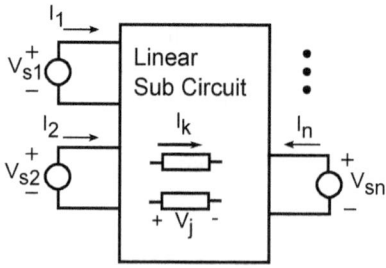

Figure 4.5: Many sources circuit

The principle can be stated as follows:

4.3. SUPERPOSITION THEOREM

Superposition: Let a circuit have n independent sources $z_1, z_2, \ldots, z_n$. Any voltage V_j or current I_k in the circuit, including voltage or currents in the sources may be expressed in the form

$$V_j = V_{j1} + V_{j2} + \ldots + V_{jn} \tag{4.2}$$

and

$$I_j = I_{j1} + I_{j2} + \ldots + I_{jn} \tag{4.3}$$

where V_{jh} is the value of voltage V_j and I_{kh} the value current I_k when all independent sources except Z_h, $h = 1, 2, \ldots, n$ are 0 (turned-off)

Fig. 4.6 illustrates the superposition theorem by steps. Although the theorem is valid for any linear subcircuit, the restriction to resistive circuits is enhanced to illustrate the principle below of interest to us.

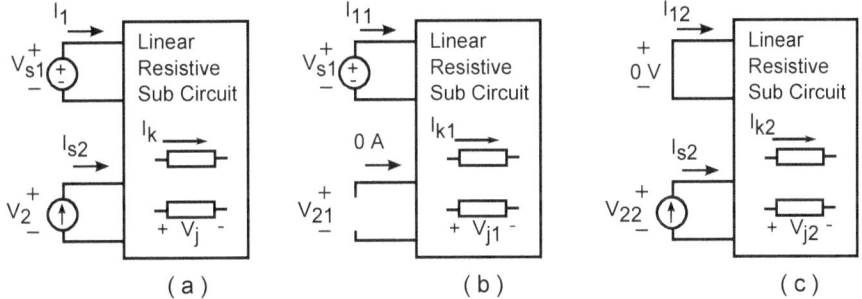

Figure 4.6: Many sources circuit

Notice that the individual calculations illustrated in steps (b) and (c) work with only one source. Hence, the properties mentioned in section 4.2 apply. In particular, we can extend the theorem of homogeneity (and proportionality) as follows:

Let x_j be a current or a voltage in a linear resistive circuit with sources $z_1, z_2, \ldots, z_m$, with

$$x_j = x_{j1} + x_{j2} + \ldots + x_{jm}$$

where x_{jh} is the value that results when all sources except z_h are off. Then, for the set of values $\{\, k_1\, z_1, k_2\, z_2, \ldots k_m\, z_m\, \}$ the response will be

$$x_j = k_1\, x_{j1} + k_2\, x_{j2} + \ldots + k_m\, x_{jm} \tag{4.4}$$

In particular, for resistive linear networks, k_j may be a function of time.

Example 4.4 A circuit has three sources, V1=2 V, I2 = 4 mA and V3 = 1.8 V. The response of interest is a current i_o. The following results are known:
- When I2 and V3 are turned off, $i_o = 2.8$ mA.
- When V1 and V3 are turned off, $i_o = 4.7$ mA.
- When V1 and I2 are turned off, $i_o = 6.5$ mA.

1. What is the value of i_o?

2. How do you express i_o as a function of the three sources?

3. What is the individual contribution of source to i_o if V1=3.6 V, I2 = 11.3 mA and V3 = 7.3 V? What is the the total value of i_o in this case?

Solution: *Let us set engineering display The first question can be answered by simply adding the individual values. Yet, to explore further the different questions, let us create lists:*

For i_o: {2.8 [EE] [(-)]3, 4.7 [EE] [(-)]3, 6.5 [EE] [(-)]3} [STO▶] X

For sources V1, I2 *and* V3, *in that order:* {2, 4.[EE] [(-)]3, 1.8} [STO▶] Y

Let us now look at the answers:

1. *Since i_o is the sum of partial answers:* sum(X) → 14.E-3 *tells us that i_0 =14 mA*

2. *The question is answered by assuming that all sources are equal to 1, and scaling the individual contributions. That is, divide each contribution by the respective source value:*

X/Y → {1.4E-3, 1.175E0, 3.611E-3}

which means

$$i_o = 1.4 \times 10^{-3} V_1 + 1.175 I_2 + 3.611 \times 10^{-3} V_3$$

3. *The first question of this item can be answered by scaling i_o contributions to the new source values. This is similar to substitute the new values in the expression of the previous item:*

{3.6, 11.3 [EE] [(-)]3, 7.3} × X/Y → {5.04E-3, 13.278E-3, 26.361E-3}

The total value of i_o is found with

sum(ans(1)) → 44.68E-3

and therefore i_o = 44.68 mA

4.3. SUPERPOSITION THEOREM

The superposition theorem can be applied to yield results faster for situations such as calculating Thevenin and Norton equivalent circuits, two-port and multi-port parameters. To extend two-port or multiport parameters methods to circuits containing independent sources, and so on. This is why a whole chapter is devoted to this property.

When combined with other theorems, like the substitution theorem, it provides further tools to deal with circuits containing reactive elements in time domain, non linear elements, etc. These applications fall outside the scope of this book but will be introduced in a second volume.

Writing equations for superposition

Lists and matrices make it easier to work superposition in few steps. This has already been illustrated with the previous example. Other examples may be found in section 5.3.1 on page 71. Let us introduce the general way of how we set up equations for methods solving equations with matrices. This is done next using the concepts introduced in section 3.5.2 on page 37.

It can be shown that any equation of the circuit can be writen as

$$a_{i1}\, x_1 + a_{i2}\, x_2 + + \ldots + a_{im}\, x_m = b_{i1}\, z_1 + b_{i2}\, z_2 + + \ldots + b_{in}\, z_n$$

All equations can be written then in matrix form as

$$\mathbf{A}\,\mathbf{x} = \mathbf{B}^{(1)}\, z_1 + \mathbf{B}^{(2)}\, z_2 + \ldots + \mathbf{B}^{(n)}\, z_n = \mathbf{B}\,\mathbf{z} \qquad (4.5)$$

where $\mathbf{A}$ is the coefficient matrix of order $m \times m$, $\mathbf{B}^{(j)}$ is a column vector of order $m \times 1$.

Notice that if all z's are zero, except z_h, then (4.5) reduces to

$$\mathbf{A}\,\mathbf{x}_h = \mathbf{B}^{(h)}\, z_h \qquad (4.6)$$

From the practical point of view of numerical calculations, we see that solving (4.5) we have

$$\mathbf{x} = \mathbf{A}^{-1}\,\mathbf{B}\,\mathbf{z} \qquad (4.7)$$

In this expression, $\mathbf{A}^{-1}\,\mathbf{B}$ has all the information concerning the individual contributions of the sources and thus what we need to obtain the desired results is there. The j-th column shows the contribution of source z_j.

Hence, when writing down equations, instead of putting all terms in one column, write the set in the form (4.7). In your calculator all you need is $\mathbf{B}$, with the vector $\mathbf{z}$ used for you to interpret results. Different examples of application are illustrated along the book.

4.4 Closing remarks

There are of course more theorems that those presented in this chapter, and useful to develop methods for working with calculators. Those in this chapter were specifically mentioned because they are applied throughout the entire book.

As the reader looks at examples of applications of theorems, awareness of the importance of understanding and using theorems will become more evident. Not only because of the easiness they may introduce in setting up algorithms for solution, but also because they provide us with the necessary tools to understand and broaden the meaning of results.

CHAPTER 5

Analysis by transformations and reduction

The goal of this chapter is to review and provide examples for several formulas associated to subcircuits used in the reduction and transformation of circuits. In this group we have the series and parallel connections, voltage and current dividers, elementary operational amplifiers structures and so on.

Since we are using calculators, we include formulas with resistances and conductances. Remember that if the datum is a resistance R, the conductance is applied as $1/R$ or R^{-1}.

The methods used in this chapter rely on intuition, and are indeed very important for the student both as a professional engineer as well as developing insight into design procedures. Of course, alternative ways to solve a problem are always possible. The solutions and methods presented try to emphasize the use of calculator or to highlight a particular feature of the calculator.

5.1 Series and Parallel Connections

The most basic reduction formulas are those for the equivalent resistance of series and parallel sub circuits (Figure 5.1). Let us start then with these cases.

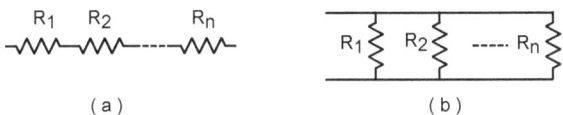

Figure 5.1: (a) Series connection of Resistances; (b) Parallel Connection of Resistances

5.1.1 Equivalent resistance and conductance formulas

For Series Connection: $R_{eq} = R_1 + R_2 + \ldots + R_n = \sum_{j=1}^{n} R_j$ (5.1a)

For Parallel Connection: $R_{eq} = \left(\dfrac{1}{R_1} + \dfrac{1}{R_2} + \ldots + \dfrac{1}{R_n} \right)^{-1}$ (5.1b)

The equivalent conductances in terms of resistances follow directly too. Remember that $G_{eq} = 1/R_{eq}$.

For Series Connection: $G_{eq} = \dfrac{1}{R_1 + R_2 + \ldots + R_n}$ (5.2a)

For Parallel Connection: $G_{eq} = \dfrac{1}{R_1} + \dfrac{1}{R_2} + \ldots + \dfrac{1}{R_n} = \sum_{j=1}^{n} \dfrac{1}{R_j}$ (5.2b)

For the particular case of two parallel resistances, (5.1b) can be reduced to the popular formula

For Parallel Connection of two resistances: $R_{eq} = \dfrac{R_1 R_2}{R_1 + R_2}$ (5.3)

This form is preferred by most students for two reasons. First, data is usually given in resistance values. Second, this form involves only one division, while the deployed one involves three. For hand analysis, these are good reasons. But when working with calculators, this should not be quite a problem. Let us illustrate this point with an example.

Example 5.1 *Fig. 5.2 shows four connections of different complexity. Let us use the above formulas to find the respective equivalent resistances.*

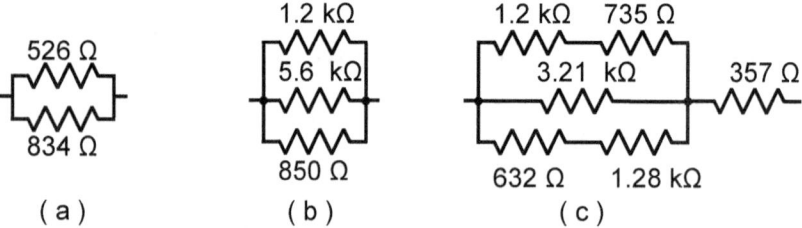

Figure 5.2: Some exampes for Series and Parallel connections

Let us work each case with separately. For this first example, I will show in particular the sequence of keys to be pressed. Later, I will limit myself to show mainly what appears on the command line. Remember that any expression of the form 1/N (N a number) is entered as 1 ÷ N, but can also be entered as N ∧

5.1. SERIES AND PARALLEL CONNECTIONS

[(-)] 1.

CASE a) *Resistances* 526 Ω *and* 834 Ω *are in parallel* (526||834)[1]. *We find the equivalent using* (5.1b) *and enter*

(1/526+1/834)∧ [(-)] 1 [ENTER] → 322.56, *pressing 17 keys.*

Notice that we could have entered

1/(1/526+1/834) [ENTER] → 322.56   *for 15 keys.*

On the other hand, (5.3) *requires 19 keystrokes, which is very comparable. Let's see now the other cases.*

CASE b) *Let us now calculate the second connection,* 1.2 kΩ||5.6 kΩ||850 Ω. *For this example, I will use again* (5.1b) *which is now much faster than successive applications of* (5.3) *when using the calculator –25 keystrokes against 40.–*

1/(1/1.2E3+1/5.6E3+1/850) → 456.96

Remarks: Remember that the E *symbol expresses a power of 10, and is entered with the* [EE] *key.*

Expression (5.1b) *has an advantage: it allows us to also see the value of the conductance. To show this, we recalculate the equivalent resistance using an intermediate step without changing the number of keystrokes (except for an additional* [ENTER]*):*

1/1.2E3+1/5.6E3+1/850 → 0.00219

Thus, the conductance is 2.19 mS. *To get the resistance, now press*

[∧] [(-)] [1]

to automatically introduce introduce the ANS(1) *variable, and the result is*

ans(1) ∧⁻1 → 456.96

Important remark: Remember that if you start your command line directly with any operation of the right column of keys (([∧], [÷], [×], [-] or [+]) or the storage key [STO▶] *, the previous result is automatically entered and appears on the entry line as* ans(1).

CASE c) *Now for the series connection of* 357Ω *with the parallel connection of resistances,* (1.2kΩ+735Ω), 3.21kΩ *and* (632Ω+1.28 kΩ) *we first calculate the*

---

[1]We write R1 || R2|| R3 ||... to denote a parallel connection of resistances R1, R2, R3, ...

conductance of the parallel connection. To reduce the risk of false keystrokes, it is better to do it by steps, using the property of ans(1) mentioned in the remark. In addition, let us work in Engineering floating 5 modes. Do:

1/(1.2E3+735)+1/3.21E3+1/(632+1.28E3) → 1.3513E⁻3
ans(1) ∧⁻1 + 357 → 1.097E3 or ANS⁻¹ + 357 → 1.097E3

Hence, the equivalent resistance is 1.1 kΩ, rounding to one digit in the decimal fraction.

Scaling units Since you are the one using the calculator, you may accelerate your work if you adopt the habit of using the appropriate units. For example, by adopting kilo ohms instead of ohms, you can save keystrokes. Your results for resistances are interpreted in kilo ohms while those of conductances as milli siemens. Thus the two last lines would be entered as

1/(1.2 + .735)+1/3.21+1/(.632+1.28) → 1.3513 (mS)
ans(1) ∧⁻1 + .357 → 1.097 (kΩ)

5.1.2 Another example of Series-Parallel Reduction

Let us look at an example which is very "hand-like", in the sense that we follow more or less the same steps as those in a hand analysis, except perhaps for the use of the operations. This type of examples appears in the many situations such as when step by step considerations are required or when a compact plan is not so direct.

Example 5.2 *Let us find the equivalent resistance of the circuit on Figure 5.3 by reducing it to only one resistance, namely, the equivalent resistance. We present two decimal figures in results.*

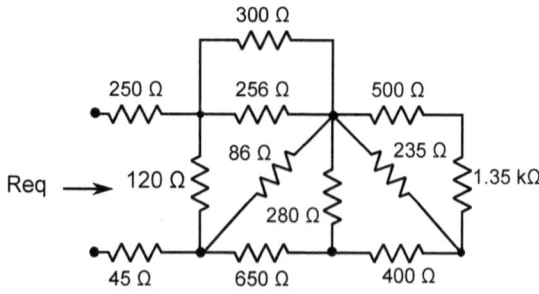

Figure 5.3: Example for series-parallel reduction

5.1. SERIES AND PARALLEL CONNECTIONS

We first identify at the right side of the circuit 400 Ω in series with the parallel connection of 235 Ω and the series connection 500 Ω- 1350 Ω. Using the formula for parallel connection, while including the series formula within the same expression, we have

$$400 + \left(\frac{1}{235} + \frac{1}{500 + 1350}\right)^{-1} = 608.51$$

which is entered as

400 + (1÷ 235 + 1÷ (500 + 1350)) ∧ (-) 1 ENTER → 608.51

The process is shown in Figure 5.4. In this figure I highlight the fact that the result is now in stack as ANS(1) in the TI-89. For simplicity, the figure shows [ANS] only.

Remember! In hand analysis you would probably have preferred the classical R1R2/(R1+R2) reduction, which is correct. But now we take advantage of calculator for the numerical steps and can introduce inverses with no remorse or fear.

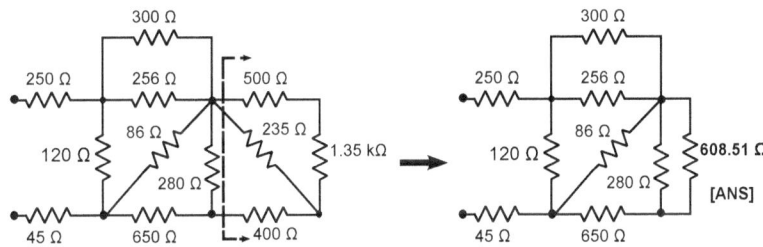

Figure 5.4: Example for series-parallel reduction, first steps

The next reduction will be done in short steps. We reduce the sub circuit in parallel with the 86 Ω resistance. This is built with a 650 Ω resistance in series with the parallel connection of 608.51 Ω –which is already on stack, as illustrated –, and 280 Ω. The reduction is done with the expression below, and the result is illustrated in Fig. 5.5.

$$650 + \left(\frac{1}{ANS} + \frac{1}{280}\right)^{-1} = 841.76$$

which in two short steps can be entered as follows:

1 ÷ ANS + 1 ÷ 280 ENTER → 5.21E-3
∧ (-) 1 + 650 ENTER → 841.76

We now reduce separately the parallel connections 841.76 || 86 Ω and 300 || 256 Ω. In the TI-89 we can use the stack to store intermediate steps; if you prefer, we can also store to a variable, say X:

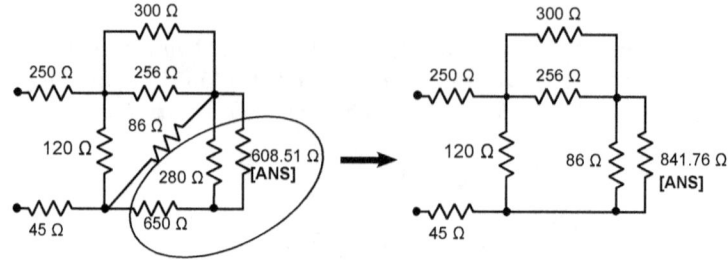

Figure 5.5: Continuing with the example for series-parallel reduction

$$\left(\frac{1}{ANS} + \frac{1}{86}\right)^{-1} = 78.03$$

which is realized in two steps as:

1 ÷ ANS + 1 ÷ 86 [ENTER] → 0.013
∧ [(-)] 1 [ENTER] → 78.03

In a similar way, in two steps, we enter

$$\left(\frac{1}{300} + \frac{1}{256}\right)^{-1} = 138.13$$

Since we have used two steps here, the result 78.03 has been pushed two steps on the stack, so it is available as ANS(3), while 138.13 in ANS(1). This intermediate step is clearly shown in Fig. 5.5. Therefore you can apply

$$ANS(1) + ANS(3) = 216.16$$

to reduce the series connection.

This result is in parallel with the resistance of 120 Ω, so we find ANS∥120 = 77.16 Ω. The final result becomes then

R_{eq} = ANS(1) + 250 + 45 = 372.16 Ω.

The step by step process to this point is illustrated in Fig. 5.6.

5.1.3 Using Lists in series-parallel

We can use lists to calculate series and parallel equivalents. Defining

$$\mathbf{R} = \{R_1, R_2, \ldots, R_n\}$$

5.1. SERIES AND PARALLEL CONNECTIONS

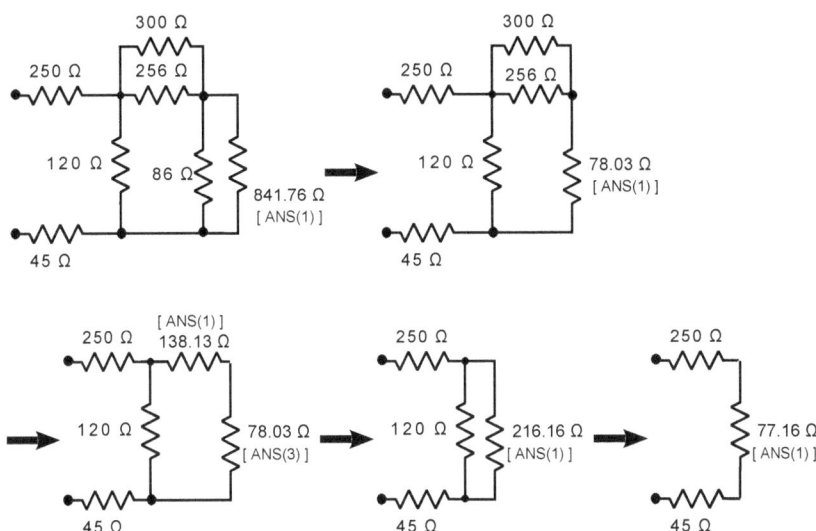

Figure 5.6: Example for series-parallel reduction

we proceed with formulas es follows:

$$\text{For Series Connection: } R_{eq} = \text{sum}(\mathbf{R}), \quad G_{eq} = \frac{1}{\text{sum}(\mathbf{R})} \quad (5.4a)$$

$$\text{For Parallel Connection: } G_{eq} = \text{sum}(1/\mathbf{R}), \quad R_{eq} = \frac{1}{\text{sum}(1/\mathbf{R})} \quad (5.4b)$$

Remember that the function sum(is obtained from the matrix menu, or may be typed directly.

The advantage of using lists is that it is less prone to mistakes and allows more complex calculations too when combined with memory or with stack. It may be or not faster, depending on particular situations. In general, though, it allows better control if you are using the list in a broader context. Let us start with the basic examples previously presented.

Example 5.3 *For Fig. 5.2(a), we define*

{526, 834} STO▶ X

to find

1/sum(1/X) → 322.56

For case (b)

{1.2E3,5.6E3,850} [STO▶] X

to find
1/sum(1/X) → 456.96

For case (c),

{1.2E3 + 735,3.21E3,632+1.28E3} [STO▶] Y

to find
1/sum(1/Y) + 357 → 1.097E3

5.1.4 Function Parallel pl(z)

You could argue that lists are not always convenient, and you may be right. However, it is easy to define a function for calculation of the equivalent resistance of a parallel combination with lists as follows [2]:

$$\text{DEFINE pl}(z) = 1 \div \text{sum}(1 \div z) \tag{5.5}$$

where z is the list of resistances in parallel.

Alternatively, we could define the function using the store key:

$$1 \div \text{sum}(1 \div z) \;\boxed{\text{STO▶}}\; \text{pl}(z) \tag{5.6}$$

One the advantage of the function definition is that you can use it directly as another value. We shall see the convenience of this function in the example below.

Example 5.4 *We want to find the equivalent resistance for the one port of Figure 5.7. All resistance values are in ohms (Ω), so the equivalent resistance will also be in ohms. Or it could be in kilohms, in which case the same units apply to the answer :)*

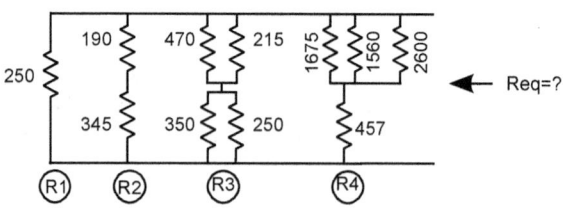

Figure 5.7: Example for Req calculation. Values are in Ω. Individual branches are denoted with circled labels.

---

[2] If you have doubt on function definition, review Section § 2.5 on page 16

5.1. SERIES AND PARALLEL CONNECTIONS

The subcircuit consists of four branches in parallel, which are identified by circled R's; branches R2, R3 and R4 are themselves series and series-parallel sub circuits. Since

$$R2 = 345 + 190$$

we can use the expression directly. Now let us work the complete subcircuit, using the stack or auxiliary variables for easiness.

First create the individual branches separately and then create the list $\mathbf{x} = \{250,$ *R2, R3, R4*$\}$. *This method is easier to correct if an error is introduced. I use the stack or auxiliary variables for R3 and R4. Execute the following steps. Here, for simplicity I limit the answer to three decimal figures:*

Step 1 *Enter R3 on stack:*

   pl({350, 250}) + pl({470, 215}) : [STO▶] Y [ENTER] → 293.352

Step 2 *Enter R4 on stack, where now R3 is pushed up:*

   457 + pl({1675, 1560, 2600}) [ENTER] → 1086.846

Step 3 *Create final list:*

   {250, 345+190, ANS(2), ANS(1)} [STO▶] X →
   {250.000 535.000 293.352 1086.846}

Step 4 *Equivalent resistance:*

   pl(ANS(1)) [ENTER] *or* 1/sum(1/ANS(1)) → 98.057

Let us take the next example using series-parallel reduction, using lists again to show the advantage of these graphical calculators. However, the important fact that I want to bring to your attention is the convenience **to plan the strategy** before using the calculator, so you can make problem solving more effective. This is important also in this particular example because this example illustrates a case where the reduction step is only part of the process. There is another part which consists in "going back" through the procedures to obtain the required answers.

This next example, on the other hand, provides a very nice application of lists. Try the example without using lists to see the difference! I have also used the previously defined function pl(z), but you may substitute it by the respective list operation if you prefer.

Example 5.5 *Find the voltage and current at each resistor in Fig. 5.8. Configure the calculator mode in* ENG. *That is, results are expressed engineering format display mode.*

Solution *Let us redraw the circuit with currents and voltages as shown in Fig. 5.9 (a). Using this notation we can now proceed with the steps to solve our problem.*

Step 1. *Since the 1.5, 2.3, and 3.8 kilo ohm resistors are in parallel, and thus with the same voltage* V_1, *we will be able to find the respective currents with Ohm's Law in one step using a list as*

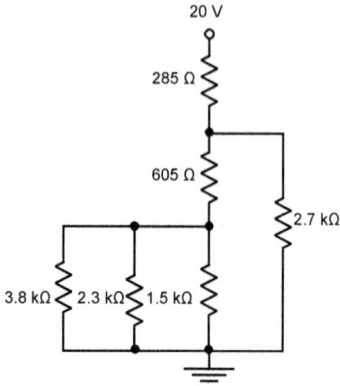

Figure 5.8: Another example with series-parallel reduction

$$\{I_1, I_2, I_3\} = V_1 / \{1.5E3, 2.3E3, 3.8E3\} \tag{5.7}$$

Therefore, our first step will be to create the list of these resistances in parallel.

Step 2. *Use the list to obtain an equivalent resistor* Rx *using eq. (5.4b) and reduce the circuit to that of Fig 5.9 (b).*

Step 3. *From this reduced circuit we see that* V_1 *and* V_4 *can be found once we have* I_4 *by using*

$$\{V_1, V_4\} = I_4 \times \{Rx, 605\} \tag{5.8}$$

Similarly, the currents in this parallel subcircuit may be calculated with

$$\{I_4, I_5\} = V_5 \div \{Rx + 605, 2700\} \tag{5.9}$$

Therefore, for this step we need to create the lists of resistances used in these equations.

Step 4. *Again, using the last list, we can reduce to an equivalent resistor* RY *arriving at Fig. 5.9 (c). Now we can see how the following equations are applied.*

$$\{V_6, V_5\} = I_6 \times \{285, Ry\} \tag{5.10}$$

so we create another list with these resistances.
We find our current I_6 *with*

$$I_6 = 20 \div (285 + Ry) \tag{5.11}$$

We are now ready to proceed. In parenthesis, on the right of the step, I have included the meaning of the result for this example. Remember that if you do not want to use the function **pl** *you may simple apply the corresponding formula.*

5.1. SERIES AND PARALLEL CONNECTIONS

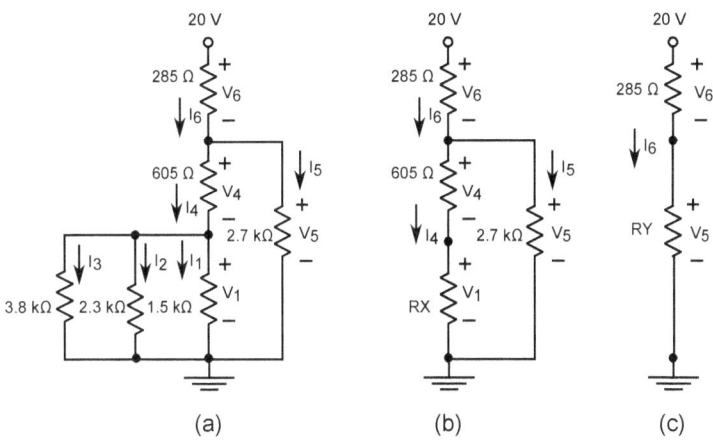

Figure 5.9: Another example with series-parallel reduction

1. Create list L1 for Step 1:
{1.5 [EE] 3, 2.3 [EE] 3, 3.8 [EE] 3} [STO▶] L1 [ENTER]

2. Calculate Rx for step 2 and store as X:
pl(L1) [STO▶] X [ENTER]   (Rx = 732.81 Ω)

3. Create list L2 for (5.9):
{X + 605, 2.7 [EE] 3} [STO▶] L2 [ENTER]

4. Calculate Ry for step 4:
pl(L2) [STO▶] Y   (Ry = 897.84 Ω).

5. Calculate I_6:
20 ÷ (285 + Y) [ENTER] → 16.909 E-3   (I_6 = 16.91 mA)

6. Calculate V_6 and V_5:
× {285, Y} [ENTER] → {4.8189E0   15.181 E0}
(V_6 = 4.82 V and V_5 = 15.18 V)

7. Calculate I_4 and I_5:
ANS(1)[2] ÷ L2 [ENTER] → {11.348E-3   5.5608E-3}
(I_4 = 11.35 mA and I_5 = 5.56 mA)

8. Calculate V_1, V_4:
ANS(1)[1] × {X, 605} [ENTER] → {8.3157   6.8654}
(V_1 = 8.32 V and V_4 = 6.87 V)

9. Calculate I_1, I_2, and I_3:
ANS(1)[1] ÷ L1 [ENTER] → {5.5438E-3   3.6155 E-3   2.1883E-3 }
(I_1 = 5.54 mA, I_2 = 3.62 mA, and I_5 = 2.19 mA)

Notice here the advantages of using lists stack for faster calculations. No need to re-enter values with the corresponding rounding errors. Moreover, see how using engineering mode makes it easier to read results in SI units.

An example with Homogeneity property

Let us take advantage of this example to see a nice application of the homogeneity principle.

Example 5.6 *For the circuit in Example 5.5 on page 57, find the voltages and currents in the circuit if the 20 V source is substituted by*
A) a voltage source of 5.6 V.
B) a voltage source of $15e^{-2t}$ V.
C) A voltage source V_s.

Solution:
The values of the currents and voltages for the original circuit are shown in the first columns of the two tables below. Let us fix the number of decimals to 2. Create lists

$$L1 = \{5.54, 3.32, 2.19, 11.35, 5.56, 16.91\} \quad \text{and} \quad V = \{8.32, 6.87, 15.18, 4.82\}$$

The second column in the tables are obtained with the factor $5.6/20$ as
5.6 $\boxed{\times}$ L1 $\boxed{\div}$ 20 $\to$ { 5.54 3.32 2.18 11.35 5.56 19.91}
5.6 $\boxed{\times}$ V $\boxed{\div}$ 20$\to$ { 8.32 1.92 4.25 1.35}

Magnitude	Original 20 V	Case 5.6 V Factor $\frac{5.6}{20}$	Case $15e^{-2t}$ V Factor $\frac{15}{20}e^{-2t}$	Symbolic V_s Factor $\frac{1}{20}V_s$
I_1 (mA)	5.54	1.55	4.16 e^{-2t}	0.28 V_s
I_2 (mA)	3.32	0.93	2.49 e^{-2t}	0.17 V_s
I_3 (mA)	2.19	0.61	1.64 e^{-2t}	0.11 V_s
I_4 (mA)	11.35	3.18	$8.51e^{-2t}$	0.57 V_s
I_5 (mA)	5.56	1.56	4.17 e^{-2t}	0.28 V_s
I_6 (mA)	16.91	4.73	12.68 e^{-2t}	0.85 V_s

Magnitude	Original 20 V	Case 5.6 V Factor $\frac{5.6}{20}$	Case $15e^{-2t}$ V Factor $\frac{15}{20}e^{-2t}$	Symbolic V_s Factor $\frac{1}{20}V_s$
V_1 (V)	8.32	2.33	$6.24e^{-2t}$	$0.42V_s$
V_4 (V)	6.87	1.92	$5.15e^{-2t}$	$0.34V_s$
V_5 (V)	15.18	4.25	$11.39e^{-2t}$	$0.76V_s$
V_6 (V)	4.82	1.35	$3.62e^{-2t}$	$0.24V_s$

The coefficients of the other two columns are obtained by multiplying lists $L1$ and V by $5/20$ and $1/20$, respectively. The factors e^{-2t} and V_s follow directly from the homogeneity principle and do not need to be included in the calculator work.

5.2 Delta-Wye and Wye-Delta transformations

Let us close dealing only with resistances by working the delta-wye (Δ to Y) and wye-delta (Y to Δ) transformations, which do not reduce the number of elements, but change the circuit topology. These transformations are specially important when working with complex impedances. Remember, this section in particular, and all the chapter too, can be worked with complex numbers when using phasors.

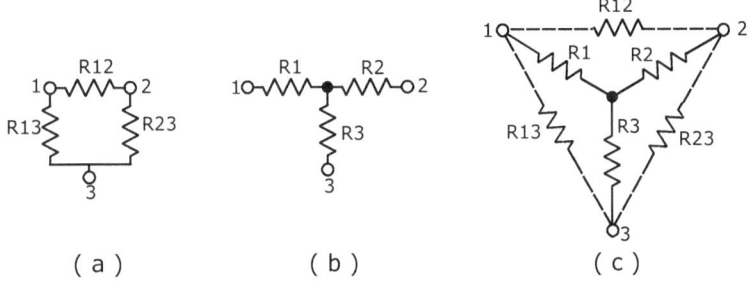

Figure 5.10: (a) Delta or Pi configuration; (b) Wye or T configuration; (c) Mnemotecnic reference

The delta - Δ - and wye - Y - subcircuits, shown in Fig. 5.10 a-b, are two useful topologies used in many situations. They are also called pi and T subcircuits, respectively. These two subcircuits are equivalent when voltages V_{12}, V_{13}, and V_{23} on one side, and currents entering into terminals 1, 2 and 3 from outside are identical. This means that from an external point of view, one can substitute one subcircuit by the other in mathematical procedures. The equivalences are also used in circuits to take advantage physically, like for example realizing high valued resistances in integrated circuits, using lower valued ones.

The equivalence of both subnetworks is satisfied when the following transformations apply.

Delta-to-Wye Transformations:

$$R_1 = \frac{R_{12}\,R_{13}}{R_{12}+R_{13}+R_{23}} \qquad (5.12a)$$

$$R_2 = \frac{R_{12}\,R_{23}}{R_{12}+R_{13}+R_{23}} \qquad (5.12b)$$

$$R_3 = \frac{R_{13}\,R_{23}}{R_{12}+R_{13}+R_{23}} \qquad (5.12c)$$

Wye-to-Delta Transformations:

$$R_{12} = \frac{R_1\,R_2+R_1\,R_3+R_2\,R_3}{R_3} \qquad (5.13a)$$

$$R_{13} = \frac{R_1 R_2 + R_1 R_3 + R_2 R_3}{R_2} \tag{5.13b}$$

$$R_{23} = \frac{R_1 R_2 + R_1 R_3 + R_2 R_3}{R_1} \tag{5.13c}$$

Fig. 5.10c provides a reference view of the formulas. In the first set of formulas, notice that the numerator is the product of the two resistances of the Delta structure that are connected to the same terminal of the Y-structure resistance. In the second set, the denominator is the non connected resistance of the Y-structure to the respective Delta-structure resistance.

Most textbooks use Ra, Rb, and Rc for the Delta resistances. I prefer to use double subscripts because then the formulas are easier to remember. In the first set of equations, the common subscript in the resistances of the numerator gives the subscript for the result. In the second set, the subscript of the Y resistance is the one not present in the Delta resistance of the result.

5.2.1 Working formulas directly

You can work formulas (5.12) and (5.13). Use memory/stack for easiness. You may combine with lists too.

Example 5.7 Delta to Wye: *Take a Delta subcircuit with resistances* $R_{12} = 651$ Ω, $R_{13} = 1200$ Ω, $R_{23} = 760$ Ω. *We proceed to calculate with formulas (5.12)*

Step 1- Store Denominator: 651 + 1200 + 760 → 2611

Step2- Calculate Wye resistances: {651*1200, 651*760, 760*1200}/ans(1)
   → {299.2   189.49   349.29}
   Interpret result as {R1, R2, R3}

Wye to Delta: *Take now a Y subcircuit with* $R_1 = 600$ Ω, $R_2 = 313$ Ω, $R_3 = 728$ Ω. *To calculate formulas (5.13):*

Step 1- Store Numerator: 600 * 313 + 600 * 728 + 313*728 → 8.5246E5

Step2- Calculate Delta resistances: ans(1)/{728,313,600} →
   {1171   2723.5   1420.8}
   Interpret result as {R12, R13, R23}

5.2. DELTA-WYE AND WYE-DELTA TRANSFORMATIONS

5.2.2 Using functions

Let us create the function for Delta-Wye transformation for the TI-89 and the TI-:

$$\text{DEFINE pi2y(z)} = \{\text{z[1]*z[2], z[1]*z[3],z[2]*z[3]}\} \div \text{sum(z)} \quad (5.14)$$

where z is the list of Δ-resistances $\{R_{12}, R_{13}, R_{23}\}$. The output of the function is the Wye-resistance list $\{R_1, R_2, R_3\}$.

Similarly, create the function for Wye-Delta transformation:

$$\text{DEFINE y2pi(z)} = (\text{z[1]*z[2]+ z[1]*z[3]+z[2]*z[3]}) \div \{\text{z[3],z[2],z[1]}\} \quad (5.15)$$

where z is the list of Wye-resistances $\{R_1, R_2, R_3\}$. The output of the function is the Delta-resistance list $\{R_{12}, R_{13}, R_{23}\}$.

Let us look at the example.

Example 5.8 *For example 5.7:*

Delta to Wye: pi2y({651,1200,760}) → {299.2   189.49   349.29}

Wye to Delta: y2pi({600,313,728}) → {1171   2723.5   1420.8}

One advantage of using lists is the economy in keystrokes when you have defined the list before calling the function. This is convenient, for example, when the original resistances are in fact equivalent resistances of another connection, as illustrated in the following example.

Example 5.9 *Find the Y equivalent for the delta sub circuit in Fig.5.11(a). The delta resistances are in fact equivalents of series-parallel subcircuits.*

Solution: *For this example, each of the resistances to be used in the formulas is in fact the equivalent resistance of a subcircuit. Specifically, using as reference Fig. 5.10, it is seen that*
- *R12 is 4.20 kΩ in series with the parallel connection of 1 kΩ and 1.6 kΩ,*
- *R13 is the parallel connection of 2.5 kΩ with the series sub circuit of 1.3 kΩ and 13 kΩ, and*
- *R23 is the series connection of 5.61 kΩ and 1.4 kΩ.*

Having identified these equivalencies, we introduce the equivalent resistances in the first step of transformation Delta-to-Wye using lists:

{(4.2E3 + (1/ 1E3 + 1/ 1.6E3)∧ (-) 1), (1 / 2.5E3 + 1/(1.3E3 + 13E3))∧ (-) 1, (5.61E3+1.4E3)} STO▶ L6 → {4.81E3, 2.13E3, 7.01E3}

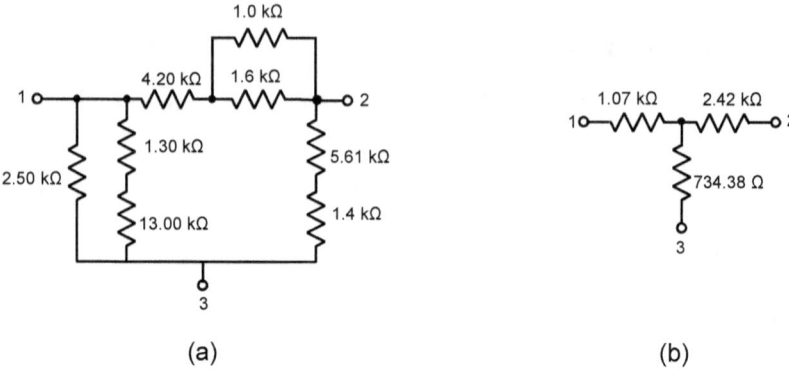

Figure 5.11: (a) Delta or Pi configuration to be transformed; (b) Equivalent Y configuration

The Y-circuit elements are obtained from

`pi2y(L6)` → $\{1.07\text{E}3,\ 2.42\text{E}3,\ 734.38\ \}$

The result should be interpreted as {R1, R2, R3 }, the configuration being that of Fig. 5.11(b).

If you prefer not to use lists in your definition, alternative ways are to explicitly name the resistances in the definition, without list notation, as shown below. Since the TI-89 has reserved the variables R12, R23 and R13 for internal use, we use letter x instead of r. Notice that to get all the resistances in one step we use a list in the definition.

Our delta to wye function (or pi to T) is defined by

$$\{\text{x12}*\text{x13},\text{x12}*\text{x23},\text{x13}*\text{x23}\}/(\text{x12 + x13 +x23})\ \boxed{\text{STO}\blacktriangleright}\ \text{pi2t(x12,x13,x23)} \quad (5.16)$$

The output from this function should be interpreted as the list { R1, R2, R3} in the Wye configuration.

Now, for the wye to delta (or T to pi) we can use

$$\text{DEFINE t2pi(x1,x2,x3)} = (\text{x1}*\text{x2}+\text{x1}*\text{x3}+\text{x2}*\text{x3})/\{\text{x3,x2,x1}\} \quad (5.17)$$

The output from this function should be interpreted as the list { R12, R13, R23} of the delta configuration.

The following example uses this notation. I also use the `pl(z)` function defined before.

5.2. DELTA-WYE AND WYE-DELTA TRANSFORMATIONS

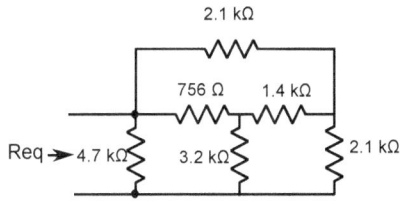

Figure 5.12: Example using Wye-Delta transformation

Example 5.10 *Find the equivalent resistance for the subcircuit in Fig. 5.12. For display, use engineering mode with four digits.*

We can identify a wye configuration with the resistances of 756 Ω, 1.4 kΩ and 3.2 kΩ. To facilitate the identification of the terminals, let us introduce labels (1), (2) and (3) as shown in Fig. 5.13(a). The steps for solution are as follows:

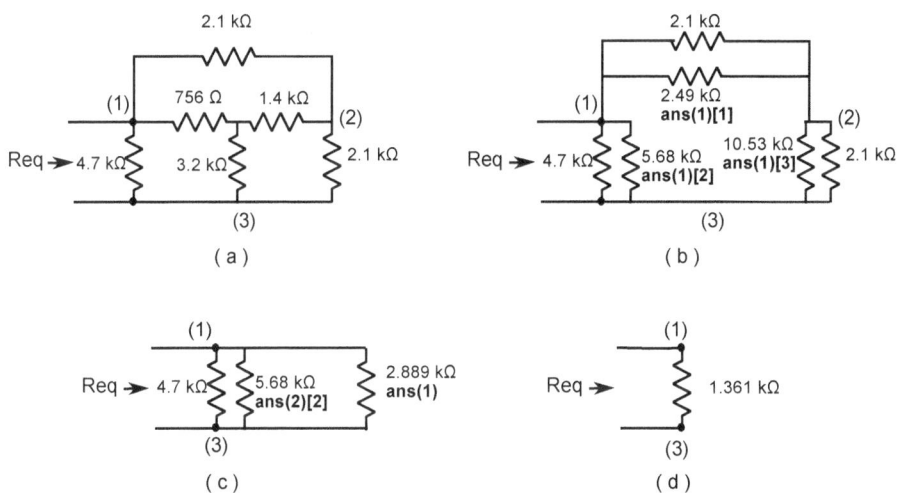

Figure 5.13: Steps in solving the example.

STEPS:	RESULT
Step 1: Apply function t2pi:	
t2pi(756,1.4E3, 3.2E3) → {2.487E3, 6.684E3, 10.53E3}	Fig. 5.13(b)
Step 2: Use parallel function pl:	
pl({2.1E3, ANS(1)[1]})+pl({2.1E3, ANS(1)[3]}) → 2.889E3	Fig. 5.13(c)
Step 3: Apply function pl again:	
pl({4.7E3, ANS(2)[2], ANS(1)} → 1.361E3	Final

5.2.3 Programming the transformations

Not everyone likes functions, many prefer programs, which are called and generate the variables directly stored as desired. Besides, results can also be displayed. The only disadvantage when displaying result is that the calculator does not necessarily take you back to the HOME environment.

Figs. 5.14 and 5.15 show two examples of programs for transformation between the two structures. The left column of line numbers is not part of the program, they are shown for reference. Also, the right column is included for documentation in this book, but is not part of the program. The program actually stored in memory is written in typewriter font.

The first program is for delta-wye transformation and is interactive, that is, it requests the input from the user. The second one for the wye-delta transformation requires the user to introduce the input values as parameters, as in the function case. The programs do not use lists because the individual values are desired in individual variables, not as elements in the list.

#	Code	Comments
1	`:delta2wye()`	
2	`:Pgrm`	
3	`:Local s`	
4	`:Input ''enter r12:    '', x12`	Requesting input
5	`:Input ''enter r13:    '', x13`	
6	`:Input ''enter r23:    '', x23`	
7	`:x13 + x12 + x23 → s`	sum in denominator
8	`:x12*x13/s→ x1`	Finding Y's resistances
9	`:x12*x23/s→ x2`	
10	`:x13*x23/s→ x3`	
11	`:Disp ''r1 ='', x1`	Displaying results.
12	`:Disp ''r2 ='', x2`	
13	`:Disp ''r3 ='', x3`	
14	`:EndPrgm`	

Figure 5.14: Delta-Wye Transformation Program without passing parameters

To run the Delta-Wye Transformation Program:

1. Type `delta2wye()` on the home command entry line
   - The IO window environment opens, asking for inputs
   - Results are displayed

2. Press HOME to exit IO window. Variables x1, x2 and x3 are available.

To run the Delta-Wye Transformation Program:

1. Type `wye2delta(x1,x2,x3)` on the home command entry line, using for x1, x2, and x3 the values for the wye structure resistances.

5.3. VOLTAGE AND CURRENT DIVIDERS   67

```
1    :wye2delta(x1, x2, x3)          Comments
2    :Pgrm
3    :Local s
4    :x1*x3 + x1*x2 + x2*x3 → s      sum in numerator
5    :s/x1→ x23                      Finding Delta's resistances
6    :s/x2→ x13
7    :s/x3→ x12
8    :Disp ''r12 ='', x12            Displaying results.
9    :Disp ''r23 ='', x23
10   :Disp ''r13 ='', x13
11   :EndPrgm
```

Figure 5.15: Wye-Delta Transformation Program passing parameters

- The IO window environment opens, displaying results

2. Press HOME to exit IO window. Variables x12, x13 and x23 are available.

Eliminating lines 8, 9 and 10 in the wye2delta progam, you will always stay at the HOME window. In both programs it is valid to provide mathematical expressions, or variables with values assigned. You can also provide undefined variables as inputs, in which case the results become functions of those undefined variables.

5.3 Voltage and Current Dividers

Other configurations of great importance are the voltage and current dividers. Fig. 5.16(a) shows the usual two-resistor divider presented in most textbooks.

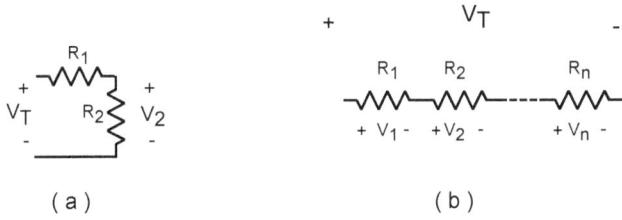

Figure 5.16: (a) Two-resistor voltage divider; (b) n-resistor voltage divider

The voltage at R_2 is given by

$$V_2 = \frac{R_2 V_T}{R_1 + R_2} \qquad (5.18)$$

In the general n-resistor case shown in Fig. 5.16b, the voltage at any resistor R_j in the series string is given by

$$V_j = \frac{R_j V_T}{R_1 + R_2 + \ldots + R_n} \tag{5.19}$$

It is possible to immediately extend this formula with lists so we can calculate all the resistance voltages in one step. Defining the list $\mathbf{R} = \{R_1, R_2, \ldots, R_n\}$, we get

$$\{V_1, V_2, \ldots, V_n\} = V_T \times \mathbf{R} \div \text{sum}(\mathbf{R}) \tag{5.20}$$

The dual subcircuits of the previous voltage dividers are the current dividers shown in Fig. 5.17. For the two-resistor case we have

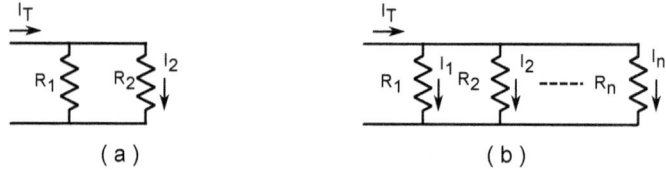

Figure 5.17: (a) Two-resistor current divider; (b) n-resistors current divider

$$I_2 = \frac{G_2 I_T}{G_1 + G_2} \tag{5.21}$$

For two resistors, the formula is more commonly worked as

$$I_2 = \frac{R_1 I_T}{R_1 + R_2} \tag{5.22}$$

Notice that the numerator involves "the other" resistance. This formula is valid just for two resistors.

For the n-resistors divider we have

$$I_j = \frac{G_j I_T}{G_1 + G_2 + \ldots + G_n} \tag{5.23}$$

As in the case of voltage divider, it is also possible to express the current divider formulas using lists and in one step obtain all the resistor currents.

$$\{I_1, I_2, \ldots, I_n\} = I_T / \mathbf{R} \div \text{sum}(1/\mathbf{R}) \tag{5.24}$$

Let us start with simple examples.

Example 5.11 *For the voltage divider in fig:VoltageDividersExample1(a), find V_1, V_2 and V_3. Also, find the currents for the current divider in (b). The calculator is*

5.3. VOLTAGE AND CURRENT DIVIDERS

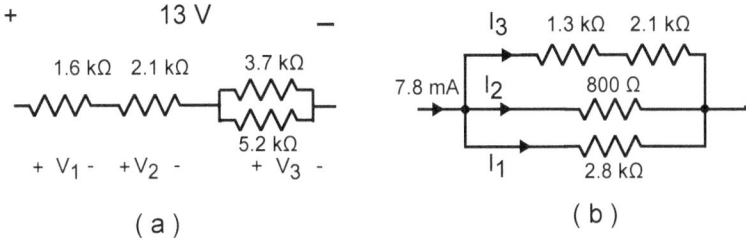

Figure 5.18: (a) Two-resistor voltage divider; (b) n-resistor voltage divider

in engineering format with two decimal figures for the results.

Solutions *(a) For the voltage divider, we go through the following steps, using list variable R for faster procedure:*

$\{1.6\text{E}3, 2.1\text{E}3, 1/(1/3.7\text{E}3 + 1/5.2\text{E}3)\}$ STO▶ R → $\{1.60\text{E}3, 2.10\text{E}3, 2.16\text{E}3\}$

$13 \times R \div \text{sum}(R)$ ENTER → $\{3.55\text{E}0, 4.66\text{E}0, 4.79\text{E}0\}$

The list that results shows the respective voltages V_1, V_2 and V_3 at the 1.6 kΩ, the 2.1 kΩ, and each of the resistances 3.7 kΩ and 5.2 kΩ in the parallel subcircuit.

(b) For the current divider, using variable R again we proceed as follows:

$\{2.8\text{E}3, 800, 1.3\text{E}3+ 2.1\text{E}3\}$ STO▶ R

$7.8 \text{ E } (-) 3 \div R \div \text{sum}(1/R)$ ENTER $\Rightarrow \{1.47\text{E}^{-}3, 5.13\text{E}^{-}3, 1.21\text{E}^{-}3\}$

Thus, the current I_1 through the 2.8 kΩ is 1.47 mA, I_2 through the 800 Ω resistor 5.13 mA, and for the series branch $I_3 = 1.21$ mA.

Remark: remember that by scaling units you may save keystrokes. In particular, using kilo ohms for resistances in this formula does not alter the division. In addition, use mA for current.

Reminder: Using the homogeneity and proportionality principles Do not forget that the numeric value given to the voltage or current sources are easy to change. In particular, if the value is symbolic, i. e. not a numerical value, you can use a value of 1 for the source and then add it as a multiplying factor in your work.

The following example shows a situation in which the list denoting the series resistance must be chosen with a specific order because of the operations involved. It illustrates, once more, that using technology in problem solving requires also a good planning. In this example I use the cumsum function for lists. This function is called with the sequence "2nd 5 3 7".

Example 5.12 *Fig. 5.19a shows a string of resistances in series used often in applications, for example in flash type analog-to-digital converters. In this subcircuit, more than focused on individual resistance voltages, interest is on the node potentials V_j. These are expressed as*

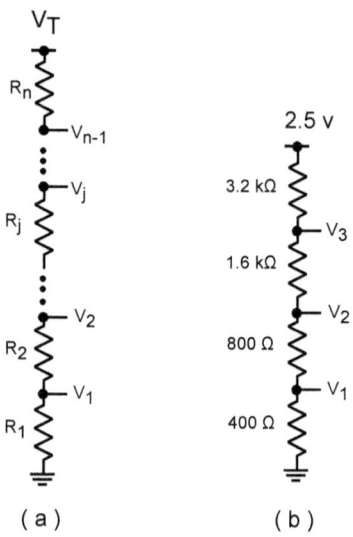

Figure 5.19: (a) A string voltage divider configuration; (b) Particular example.

$$V_j = \frac{R_1 + R_2 + \ldots + R_j}{\sum_{h=1}^{N} R_h} V_s$$

If we define $\mathbf{R} = \{R_1, R_2, \ldots, R_n\}$, *in that order,* **from ground to source**, *then we can find all potentials in one step as shown below. Observe in the formula that we include the source itself, which is the potential of node n. The reader can modify the formula to exclude it*[3].

$$\{V_1, V_2, \ldots V_n(=V_s)\} = V_s \times \text{cumsum}(\mathbf{R}) \div \text{sum}(\mathbf{R})$$

Taking the numerical example in Fig. 5.19b, and using variable X for key-stroke savings, we have:

{400,800,1600,3200 } STO▶ L1

2.5× cumsum(L1) ÷ sum(L1) enter ⇒ {.17, .50, 1.17, 2.50}

Therefore, $V_1 = 0.17$ V, $V_2 = 0.50$ V *and,* $V_3 = 1.17$ V

---

[3]See guidebook

5.3. VOLTAGE AND CURRENT DIVIDERS

5.3.1 Two-resistors dividers again: using conductances

Applications of two-resistor dividers are very large, so I thought that another look at this particular case is convenient to illustrate advantages of the calculator when the theory is applied appropriately. In this subsection, the use of conductances is stressed since this formula is better for many situations. In particular, when one of the resistances is in fact the equivalent of a parallel connection, such as in loaded dividers or passive adders.

Consider the case of two-resistor voltage divider using the formula (5.18). we can multiply numerator and denominator by $G_1 G_2$ to obtain the expression also in terms of conductances shown below.

$$V_o = \frac{R_2 V_T}{R_1 + R_2} = \frac{G_1 V_T}{G_1 + G_2} \qquad (5.25)$$

Observe that the formula can be expressed as

$$V_o = \frac{G_1}{G_{eq}} V_T$$

where G_{eq} is the equivalent conductance of the resistances in parallel.

Loaded Divider

The formula in terms of conductances allows a direct treatment in the case of a loaded divider in Fig. 5.20. Now, G_2 becomes an equivalent conductance, and we can extend (5.25) with the following two steps to calculate the output voltage using lists for any number of parallel resistors:

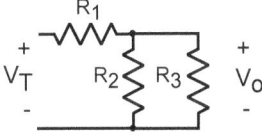

Figure 5.20: An example of a two-resistor type divider with three resistances.

$$\begin{array}{l}\text{Step 1: Define } \mathbf{G}=1/\{R_1, R_2, R_3, \ldots, R_n\} \\ \text{Step 2: Calculate } V_o = V_T \times G[1] \div \text{sum}(\mathbf{G})\end{array} \qquad (5.26)$$

Example 5.13 *If $R_1 = 350\ \Omega$, $R_2 = 950\ \Omega$, $R_3 = 1650\ \Omega$, and $V_T = 8$ V, then :*

1/{350,950,1650} [STO▶] g [ENTER] → {2.857E-3   1.053E-3   606E-6}

8 [×] g[1] / sum(g) [ENTER] → 5.062E0

Thus, Vo = 5.062 V

72 CHAPTER 5. ANALYSIS BY TRANSFORMATIONS AND REDUCTION

Example 5.14 *In the previous example, assume that the load R_3 takes the different values 200 Ω, 300 Ω, 750 Ω, 2000 Ω, and 5000 Ω. Calculate the output voltage for each case, and also the unloaded case ($R_3 = \infty, G_3 = 0$).*

Solution: *Since in this case the unloaded divider should also be considered, let us introduce the list of loads directly as a list of inductances.*

{1/200,1/300,1/750,1/2000,1/5000,0} STO▶ L3 ENTER

Then we apply (5.26) as

8* (1/350)/(1/350+1/950+L3) ENTER
→ {2.565   3.156   4.359   5.183   5.562   5.846}

The values in the final list are the respective output voltages for the different load conditions.

Let us work the example using dot operations of matrices, just for the sake of looking at something different. The procedure is a little more complicated, but the objective is to show another use of matrices.

Example 5.15 *To proceed, we follow the next steps:*

1. *Create a matrix* **G** *of all 1's, of order 3 x 6 (One column for each value of the load)*

2. *Multiply the first row by $1/R_1$, and the second row by $1/R_2$.*

3. *Substitute the third row by a row using conductances of R_3*

4. *Apply the formula of the second step in (5.26) using dot-operations:* $V_T *$ g[1]./sum(g)

Now let's go to the steps. I do not show the results on display to save space, only the last one.

1. subMat(3,6) STO▶ g: Fill 1,g ENTER

2. g[1]/350 STO▶ g[1]: g[2]/950 STO▶ g[2] ENTER

3. [1/200,1/300,1/750,1/2000,1/5000,0] STO▶ g[3] ENTER

4. 8* g[1]./sum(g) ENTER → [2.565   3.156   4.359   5.183   5.562   5.846]

The values in the final vector are the respective output voltages for the different load conditions.

Remark: *The* subMat(*and the* Fill *functions are in the matrix menu. Observe that dot operations have been used.*

5.3. VOLTAGE AND CURRENT DIVIDERS

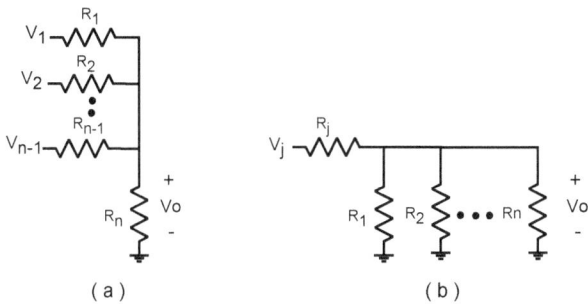

Figure 5.21: (a) A passive adder configuration; (b) When all inputs but V_j are zero (R_j not grounded)

Passive adder with voltage divider: applying superposition

Because of superposition, resistive circuits with many inputs are sometimes call adders, because they yield a weighted sum of inputs. An important case is the common passive adder shown in Fig. 5.21(a), which is basically an extension of the voltage divider.

We can apply superposition to get V_o here. When all inputs but V_j are zero, the configuration we are left with is that of Fig. 5.21(b), which is in fact a loaded divider similar to the one in Fig. 5.20. For this individual element we can write

$$V_{oj} = \frac{G_j}{G_{eq}} V_T \tag{5.27}$$

where G_{eq} is the equivalent conductance of all resistances in parallel. If you prefer to use resistances in formulas, Just for the record, this expression is sometimes expressed as

$$V_{oj} = \frac{R_{eq}}{R_j} V_T \tag{5.28}$$

The use of resistances in notation does not bring up superposition in an easy format, useful for our purposes. Hence, using (5.27) for each input, we apply superposition and get,

$$V_o = \frac{V_1 G_1 + V_2 G_2 + \ldots V_{n-1} G_{n-1}}{G_1 + G_2 + \ldots + G_{n-1} + G_n} \tag{5.29}$$

In addition to simplicity in the algebraic expression, we can work the case $R_n = \infty$, where $G_n = 0$. Moreover, since this value appears only in the denominator, there will be cases in which you simply do not need to consider it.

On the other hand, looking at (5.29), you may notice that there are $n - 1$ terms in the numerator but n in the denominator. To simplify formulas with lists, we need to comply with compatibility in dimension. Hence, the list for voltages

74 CHAPTER 5. ANALYSIS BY TRANSFORMATIONS AND REDUCTION

includes a dummy 0 V source as V_n. You can think of this dummy source to be in series with resistance R_n. If $R_n = \infty$, you may not include this dummy voltage and neither the conductance G_n; that is, both lists will be of dimension $n-1$ for $R_n = \infty$.

Let us define the two lists of resistances and sources:

1. **R**=$\{R_1, R2, R_3 \ldots R_n\}$
2. **V**=$\{V_1, V_2, V_3 \ldots V_{n-1}, 0\}$

We can now calculate equation (5.29) in one or two steps. I'll mention it as a two step process to illustrate further applications.

1. Calculate list of contributions: $\mathbf{V}_o = \dfrac{\mathbf{V} \div \mathbf{R}}{\text{sum}(1/\mathbf{R})}$ (5.30a)

2. Sum the elements of the list: `sum(Vo)` (sum(ANS(1))) (5.30b)

Notice that, <u>discarding the last element</u>, (5.30a) provides the individual contributions of the sources while (5.30b) applies superposition.

We are now in position to obtain different results easily:

A) To find the individual weights Use all sources as 1 Volt, so the individual weights are found as

$$V_o = 1 \div \mathbf{R} \div \text{sum}(1/\mathbf{R}) \qquad (5.31)$$

Discard the last element of the list for finite R_n, .

This is in particular useful when symbolic sources are present.

B) To find individual contributions When sources are of the form $k_j f_j(t)$, use the coefficients to define list **V**, and use (5.30a) to find individual contributions, after discarding the last element in he list for finite R_n. Express the result using superposition.

C) To Find the total output for numerical inputs That is, to calculate (5.29), you may either follow the two steps above (which have the advantage of providing individual contributions) or else do it in one step as

$$V_o = \text{sum}(\mathbf{V} \div \mathbf{R}) \div \text{sum}(1/\mathbf{R}) \qquad (5.32)$$

Let us take an example to look at the different alternatives.

Example 5.16 *In Fig. 5.21(a), consider n=3, with R_1=210 Ω, $R_2 = 150$ Ω, and $R_3 = 2$ kΩ. and two inputs V_1 and V_2 connected to R_1 and R_2, respectively. For this adder,*

1. *Express the ouput as the weighted sum of inputs.*

5.3. VOLTAGE AND CURRENT DIVIDERS

2. Find the individual contributions and the total output voltage Vo when the inputs are $V_1 = 2.1$ V, $V_2 = 1.8$ V.

3. Find Vo if the inputs are $V_1 = 2.1 \cos(250t)$ V, $V_2 = 1.8\, e^{-100t}$ V

Solution: For all items, we need the resistance list. Let us store the list to R

{210, 150, 2E3 } STO▶ r

The solutions follow next:

1. Enter

   1/r/sum(1/r) → {0.399   0.589   0.042}

   Discard the last element and interpret the list as $V_o = 0.399\, V_1 + 0.589\, V_2$.

2. Create list for the inputs V, {2.1, 1.8, 0 } STO▶ v

   For the individual contributions:

   v/r/sum(1/r) → {0.838   1.006   0}.

   Hence, discarding last item, the individual contributions for the 2.1 V and 1.8 V sources are, respectively, 0.838 V and 1.006 V

   Enter now sum(ANS) → 1.844   so the output is $V_o = 1.844$ V.

3. From the individual contributions in the previous item, we arrive at

$$V_o = 0.838 \cos(250t) + 1.006 e^{-100t} \quad V$$

Let us close this section with some comments about the passive adder. Many times students under appreciate "simple" circuits until they come to understand that the usefulness comes not from simplicity or complexity, but by a proper interpretation of (5.29). Let us write again this equation to look at it with a renovated approach.

Let us write the equation as

$$V_o = k_1\, V_1 + k_2\, V_2 + \ldots k_{n-1}\, V_{n-1} \tag{5.33}$$

where

$$k_j = \frac{G_j}{G_T} = \frac{R_T}{R_j} < 1; \quad \text{and} \sum_{j=1}^{n-1} k_j \leq 1$$

R_T is the equivalent resistance for all resistances in parallel. The equality sign applies when $R_n = \infty$.

Let us interpret this formula in several applications:

- **Average Value**: If $R_n = \infty$ and all resistances are equal, than Vo has the average value of all inputs.

- **DC-shift**: Again, with $R_n = \infty$, and with only two inputs, if one is time dependent, like $\cos(\omega t)$ and the other is a DC constant value source, then this source provides a controlled constant shift of the varying function.

- Summing is precisely what is done in many "mixers" used in communication systems for different applications.

To correct for the attenuation factor, the adder may be followed by an amplifier, as shown in Fig. 5.31.

To illustrate a mixer application, let us look at an "amplitude modulator" example.

Example 5.17 *This example is based on [Budak87]. In Fig. 5.21(a), take n=2 resistances, with $R_1 = R_2 =$ and two inputs $V_1 = V_m \sin(\omega t)$ V and $V_2 = V_m \sin(\omega t + \Delta \omega t)$ V connected to R_1 and R_2, respectively. We assume here that the magnitude of $\Delta \omega$ is such that both inputs are nearly equal in frequency. For this adder,*

$$V_o = \frac{1}{2}(V_m \sin(\omega t) + V_m \sin(\omega t + \Delta \omega t))$$

Based on the trigonometric identity $\sin A + \sin B = 2\cos(\frac{1}{2}[A-B])\sin(\frac{1}{2}[A+B])$ *we have*

$$V_o = V_m \cos\left(\frac{\Delta \omega}{2}t\right) \sin\left(\omega t + \frac{\Delta \omega}{2}t\right)$$

This is a sine wave of frequency $(\omega t + \frac{\Delta \omega}{2})$ *whose amplitude varies slowly at the rate* $\Delta \omega /2$. *We say that the amplitude is modulated. To see this effect, we can graph the figure on our calculator. Let us take* $V_m = 3V$, $\omega = 600$ rad/s *and* $\Delta \Omega = 50$ rad/s. *Let us put the window settings at* xmin =0, xmax = $2\pi/25$, ymin = -4 ymax = 4 *for a graph with one cycle of the envelope* $3\cos(25t)^4$. *With this settings we proceed to open Y= editor in our TI-89 and define*

```
Y1 = 1.5 sin(250*x) + 1.5 sin(300*x)
Y2 = 3 cos(25*x)
Y3 =  (-)  Y2
```

where Y2 and Y3 have been defined to show the envelope in the graph. The screen shows the graph in Fig. 5.22.

---

[4] Since the frequency is 25 rad/s, then each period is $2\pi/25$ s. Hence, xmax has been set to cover one period, or cycle, of the function.

5.3. VOLTAGE AND CURRENT DIVIDERS

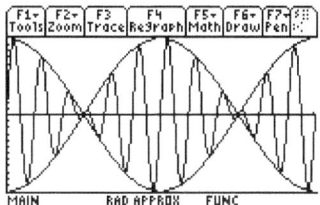

Figure 5.22: Passive adder output for the modulation example.

5.3.2 Two-resistors divider and it's Thevenin equivalent

Another useful transformation is that of the Thevenin's equivalent for a voltage divider (Fig. 5.23. The equations for the Thevenin voltage V_t and resistance R_t in this case are

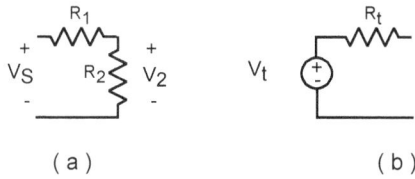

Figure 5.23: Thevenin equivalent circuit for a particular case

$$V_t = \frac{R_2 V_s}{R1 + R2} \qquad (5.34)$$

$$R_t = \frac{R_1 R_2}{R1 + R2} = R_1 \| R_2 \qquad (5.35)$$

We see that they are in fact the same as (5.25) and formula for parallel resistance, respectively. We can combine both in a function that yields $\{V_t, R_t\}$:

$$\text{DEFINE th(V,R1,R2)} = \text{R2} \times \{\text{V,R1}\} \div (\text{R1} + \text{R2}) \qquad (5.36)$$

Example 5.18 *For the ladder attenuator of Fig. 5.24, find the attenuation factor and the output resistance.*

By attenuation factor in a resistive circuit you mean the ration Vout/Vin, which will be less than 1. If you make Vin = 1 V, then the output voltage will be numerically equal to the desired factor. This means that that from the algorithmic point of view we are actually looking for is the Thevenin equivalent at the output for a voltage input of 1 V. Then V_t becomes the attenuation factor and R_t the output resistance.

CHAPTER 5. ANALYSIS BY TRANSFORMATIONS AND REDUCTION

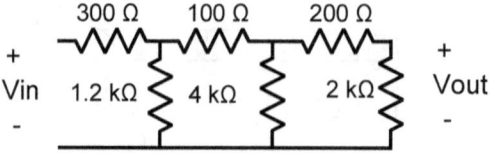

Figure 5.24: Ladder attenuator example.

The steps are shown in the table below, using the notation for the TI-89, with three decimal places, and the successive reductions in Fig. 5.25. These steps are mentioned in the right column.

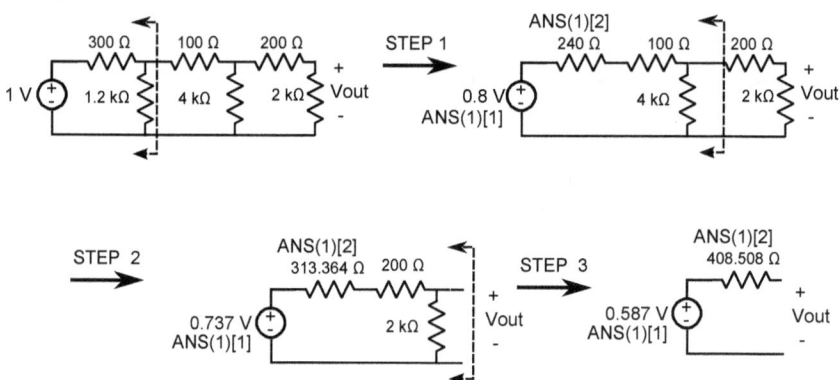

Figure 5.25: Reduction steps for the ladder attenuator example. (Stack position shown for intermediate and final values)

Command Line	Result in Stack	Reduction step
th(1,300,1200) ENTER	{0.8, 240}	Step 1
th(ans(1)[1],ans(1)[2]+100,4000) ENTER	{0.737, 313.364}	Step 2
th(ans(1)[1],ans(1)[2]+200,2000) ENTER	{0.587, 408.508}	Step 3

Therefore, the attenuation factor is 0.587 and the output resistance 408.51 Ω.

5.4 Source transformations

Source transformations, illustrated in Fig. 5.26 are very useful and common in applications. For these transformations, we apply

$$I_S = \frac{V_S}{R} \tag{5.37}$$

5.4. SOURCE TRANSFORMATIONS

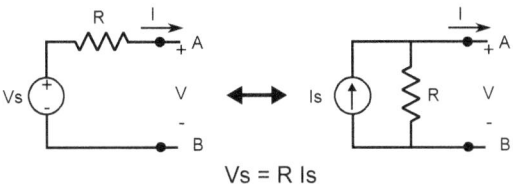

Figure 5.26: Source transformation theorem

for voltage-to-current source transformation, and

$$V_S = R I_S \qquad (5.38)$$

for current-to-voltage source transformation. Notice the great similarity with Ohm's law.

When convenient, we can realize several transformations at once using lists, as illustrated with the following example.

Example 5.19 *Consider the circuit in Fig. 5.27(a), where we want to find the current I_x. This example makes special use of the stack, but can be done the general with use of variables.*

For easiness, we use the results from the stack. The steps are illustrated in the following table and illustrated in several sub figures Fig. 5.27(b), (c), (d), and (e) respectively for each action. Fig. 5.27 also shows where in the stack – ans(1) or ans(2)– are the respective partial results. Notice the use of the pl(z) function defined in (5.5).

What is done	Entry in calculator
Convert 5V - 8Ω (left) and 8V - 14 Ω (right)	{5,8} / {8,14} ENTER → {.625 .571}
Reduce parallel sources and resistances	ans(1)+{0.5, (-) 0.3} ENTER → {1.125 .271} {pl({8,20}),pl({7,14})} ENTER → {5.714 4.667}
Convert sources and combine series resistances	ans(1) × ans(2) ENTER → {6.429 1.267} ans(2) + {6,10}) ENTER → {11.714 14.667})
Convert again	ans(2) ÷ ans(1) ENTER → {.549 .086}

We can now find the current using the current divider formula (5.23), where I_T is the sum of both sources –ans(1)–:

(sum(ans(1)/4) ÷ (1/4 + sum(1/ans(2)))) ENTER → .393

Therefore $I_x = 393$ mA. Or, looking at the screen, which shows more digits, $I_x = 393.46$ mA.

CHAPTER 5. ANALYSIS BY TRANSFORMATIONS AND REDUCTION

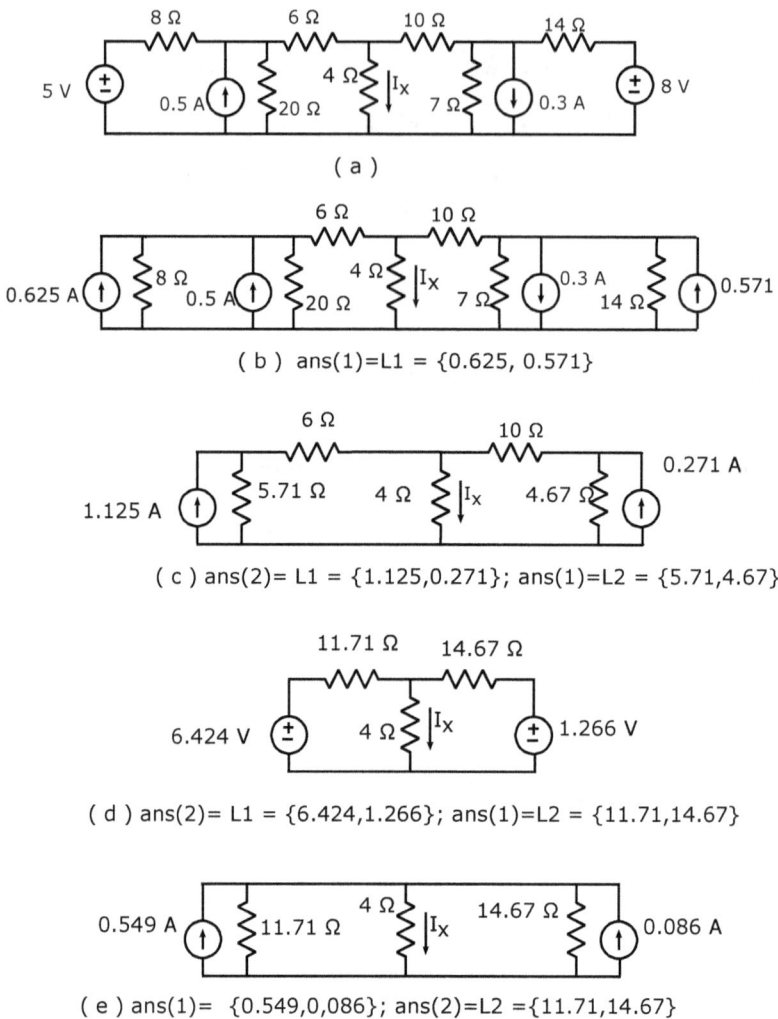

Figure 5.27: Circuit for example 5.19 on source transformation

Let us now work in general with auxiliary variables.

With intermediate variables: If "list" variables – L1 and L2– are preferred instead of stack, as it is also illustrated in the different figures of Fig. 5.27, we follow the next steps:

1. **Convert 5V - 8Ω(left) and 8V - 14 Ω(right):**
   $\{5,8\}$ / $\{8,14\}$ ENTER → $\{.625 \quad .571\}$

5.4. SOURCE TRANSFORMATIONS

[Note: Fig. 5.27(b)]

2. Reduce parallel sources and resistances:

ANS+{0.5, (-) 0.3} STO▶ L1 ENTER → {1.125   .271}

{(1/8+1/20) ∧ (-) 1,(1/7+1/14)∧ (-) 1} STO▶ L2 ENTER → {5.714   4.667}

[See Fig. 5.27(c)]

3. Convert sources and combine resistances:

ANS × L1 STO▶ L1 ENTER → {6.429   1.267}
L2 + {6,10} STO▶ L2 ENTER → {11.714   14.667}

[See Fig. 5.27(d)]

4. Convert again:

L1 ÷ L2 ENTER → {.549   .086}

[Fig. 5.27(e)]

Final Step: Find current by current division

sum(ANS(1))/4 / (1/4 + sum(1/L2)) ENTER → .393

We see once again in these steps the convenience of lists and of user defined functions to simplify procedures and speed up problem solving. Observe also that using Ohm's law with conductances instead of resistances simplified the number of steps.

5.4.1 Shifting theorems, and extended source transformations

Two theorems rarely mentioned in basic textbooks, are the shifting theorems illustrated in Fig. 5.28. Here, all currents and voltages inside subcircuit, as well as the potential at the terminal nodes, and the currents entering the subcircuit from these nodes remain unchanged. Notice that the theorems can be applied in both directions, either to split into several sources, or to reunite several onto one. Moreover, they are applicable to dependent sources.

I work now only the voltage shifting, and leave to the reader working with the dual theorem. Fig. 5.29 on page 83 shows two transformations applying the voltage shifting theorem.

In the case of (a), multiple voltage-to-current transformations are carried out at once using lists. In the second case (b), several Thevenin transformations are obtained in few steps, also using lists. The procedures are indicated in the figure themselves. The relations shown in Fig. 5.29 can be of course extended to any number of sources. Let us illustrate with one example.

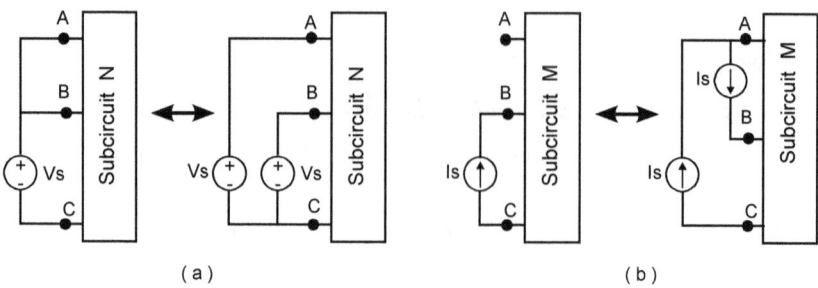

(a)   (b)

Figure 5.28: Source shifting theorems. (a) Voltage shifting theorem (b) Current shifting theorem

Example 5.20 *Consider the bridged T shown in Fig. 5.30(a), where we want to find current I. The application of the transformation shown in Fig. 5.29(b) results in Fig. 5.30(b). Calculations are as summarized in the following table:*

Entry	On stack	Comment
{300, 200} [STO▶] L1	{300. 200.}	L1 for R1
{1200, 2000} [STO▶] L2	{1200. 2000.}	L2 for R2
1 [×] L2 [÷] (L1 + L2) [STO▶] L3	{.8 .9090909}	{Va, Vb}
L1 [×] L2 [÷] (L1 + L2) [STO▶] L4	{240 181.8182}	{Rta, Rtb}
(L3[2]−L3[1]) [÷] (100 + sum(L4))	.0002091	Current

Therefore, $I = 209.1$ μA. Observe that reading would have been easier using mode ENG. Remember that sum(is entered from the list menu in you calculator.

Again, use of lists and application of theorems need practice, practice, and practice. Use different exercises from your textbooks.

5.5 Some Basic Operational Amplifier structures

The basic operational amplifier structures are the common non-inverting and inverting amplifiers. Blocks with operational amplifiers are extremely useful and have multiple applications. Remember: there are many many other structures to explore!

The next example uses the structures in Fig. 5.31. In the first two cases, when we limit the circuit to just one input, the basic structures appear. The example simply generalize the case.

Example 5.21 *Fig. 5.31 shows three common operational amplifier structures: the inverting adder, the non-inverting adder, and the adder-subtracter configurations. For simplicity, let us work with three inputs for the first two cases, and three*

5.5. SOME BASIC OPERATIONAL AMPLIFIER STRUCTURES

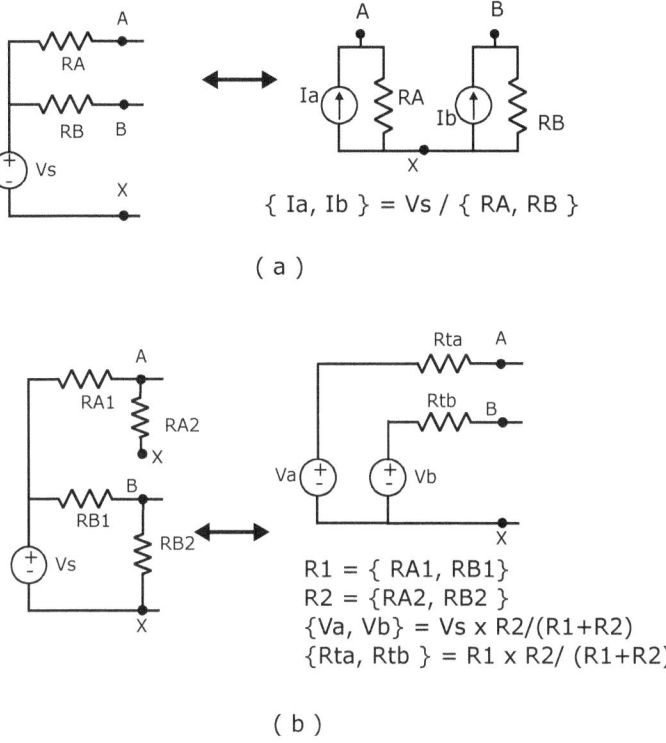

Figure 5.29: Applying source shifting theorems (a) voltage shifting to current source transformations, (b) voltage shifting for several Thevenin equivalents computation.

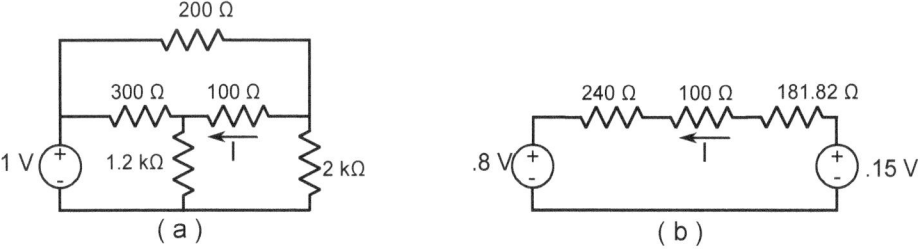

Figure 5.30: Thevenin reduction for a bridged-T.

adding voltages with three voltages to be subtracted in the third case. Equations and procedures, however, are valid for any number of cases, including the special one input amplifiers.

Let us work each case.

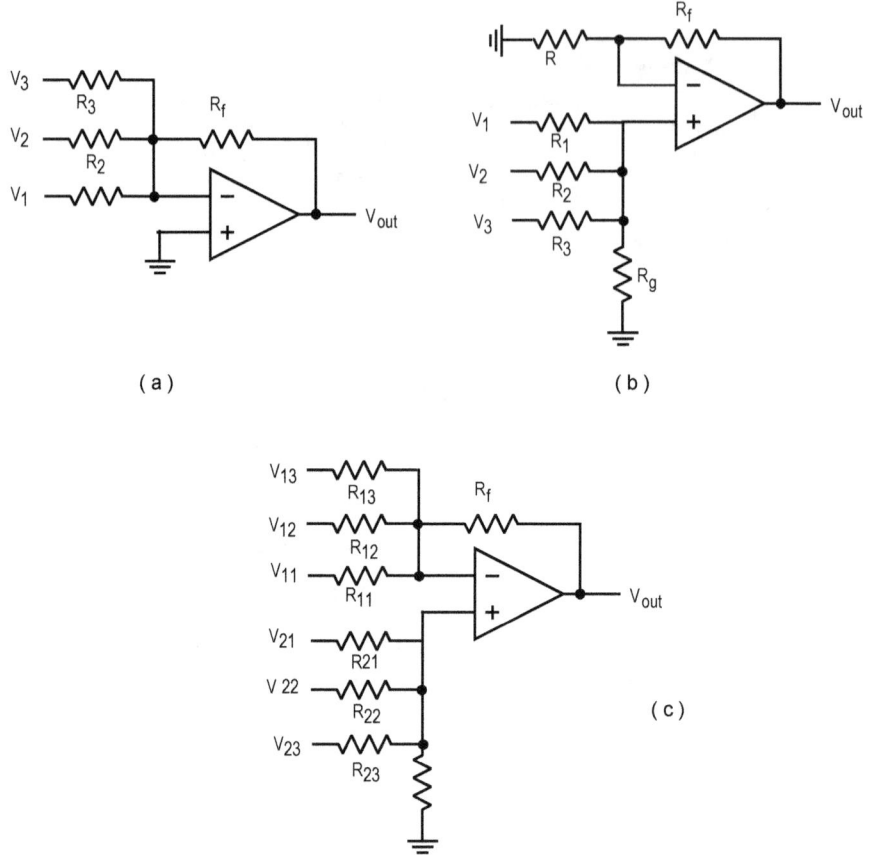

Figure 5.31: Three OA structures: (a) inverting adder; (b) non-inverting adder; (c) adder-subtracter

(A) For the inverting adder, the mathematical formula is:

$$V_{out1} = -\left(\frac{R_f V_1}{R_1} + \frac{R_f V_2}{R_2} + \frac{R_f V_3}{R_3} + \right) = R_f \sum_{j=1}^{3} \frac{V_i}{R_i} \qquad (5.39)$$

To work in the calculator, define the lists $\mathbf{R} = \{R_1, R_2, R_3\}$ and $\mathbf{V} = \{V_1, V_2, V_3\}$. Then (5.39) may be written as

$$v_{out} = -R_f \times \ sum(\mathbf{V}/\mathbf{R}) \qquad (5.40)$$

(B) The non-inverting adder is basically a passive adder with a non-inverting amplifier. Therefore, we use (5.29) multyplying it by the gain:

$$V_{out} = \frac{(1+R_f/R)}{G_1+G2+G_3+G_g}(V_1\,G_1 + V_2\,G_2 + V_3\,G_3) \tag{5.41}$$

To work this formula, define $\mathbf{R} = \{R_1, R_2, R_3, R_g\}$ and $\mathbf{V} = \{V_1, V_2, V_3, 0\}$, and then apply

$$v_{out} = (1+R_f/R) \times sum(\mathbf{V}/\mathbf{R})/sum\,(1/\mathbf{R}) \tag{5.42}$$

(C) Finally, for the adder-subtracter, we combine (5.39) and (5.41), where R is the parallel combination of $R_{11}, R_{12}, \ldots$. The sequence in the calculator, using variables (X, Y, R, A, B) for simplicity, may be as indicated below:

1. $\{R_{11}, R_{12}, R_{13}\}$ STO▶ X
2. 1÷sum(1/X) STO▶ R
3. $\{V_{11}, V_{12}, V_{13}\}$ STO▶ Y
4. (-) $R_f \times$ sum(Y/X) STO▶ A
5. $\{R_{21}, R_{22}, R_{23}, R_g\}$ STO▶ X
6. $\{V_{21}, V_{22}, V_{23}, 0\}$ STO▶ Y
7. (1 + R_f/R)× sum(Y/X)/sum(1/X) STO▶ B
8. *Output voltage is now* A + B

The reader is encouraged to try numeric values for data.

5.6 Closing Remarks

As you might have imagined by now, it is possible to stay with reduction and transformation methods for the rest of the book. These methods require from you to develop your intuition, which is excellent in many ways, both for your professional development and your improvement in analysis and tools manipulating skills. Literature is full of such methods; you are encouraged to look at them.

As a suggestion, develop a library of functions or programs, specially for those cases which appear often, so you can boost your speed in solving problems. But write the functions yourself, don't look for them at the internet, unless you already know how to do it. Otherwise, by doing it yourself you don't miss the joy of learning and discovering! Once you have your set of functions, develop your custom menus for easy use.

Finally, it is important to realize that steps and procedures depend on your intuition and, to develop the intuition you must practice, practice and practice. But don't forget: practice paying attention to procedure!

CHAPTER 6

Nodal Analysis

Nodal analysis is one of the most popular analysis methods based on systematic set up of equations. The chapter review here effective ways to work these equations taking advantage of the technology, including also sections on programming. This chapter assumes that the reader can use matrices to solve linear equations. If necessary, please review Chapter 3, section 3.2 on page 25, or else the calculator manual.

6.1 Introduction to the theory of nodal analysis

Nodal analysis uses the concept of *node potential*, which is the voltage difference between a node k and a reference node called *ground* which, by definition, has potential of 0 V. Any other voltage can be expressed in terms of these potentials, as illustrated in Fig. 6.1. A circuit with N nodes will have N-1 node potentials.

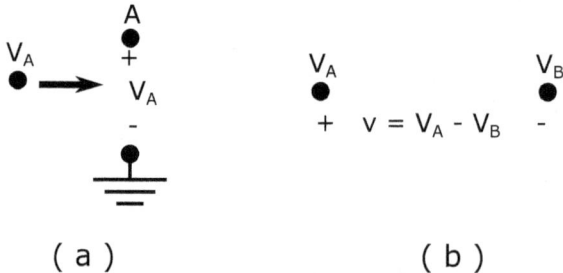

Figure 6.1: a) The concept of node potential, b) voltage between two nodes

Mathematically, any node can be chosen as ground. If you have already found the potentials for a circuit and decide to change ground to another node B, then all potentials can be adjusted to the new reference by subtracting the original potential

6.1. INTRODUCTION TO THE THEORY OF NODAL ANALYSIS

V_B, as illustrated in Fig. 6.2. Therefore, do not worry too much if you selected one node or other as reference. However, pay attention to the sign voltages.

Remark: This ground selection is of mathematical nature. Physical circuits are built with a ground in mind. This is particularly important in many cases, such as for example layouts of printed circuit boards and design of integrated circuits. But there are many other instances.

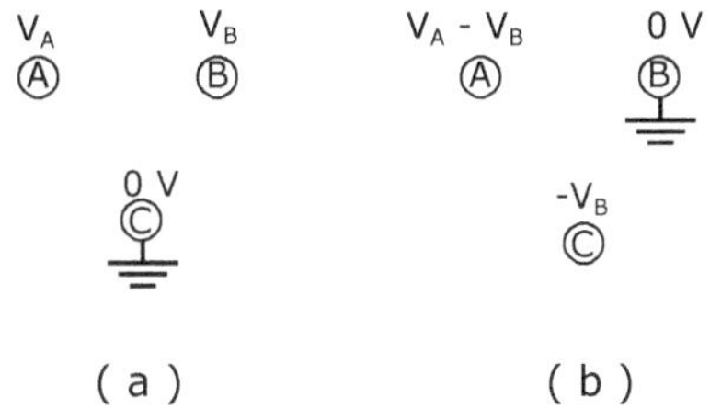

Figure 6.2: Changing reference and effect on potentials: a) original, b) after change

The nodal method consists in finding the circuit variables by setting up a set of equations, called *nodal equations*, where the node potentials are the main variables. The equations however may also include currents not dependent on voltages. Once the equations are solved for the node potentials and those currents, any other circuit variable can be calculated.

Each basic nodal equation is actually a current equation based on Kirchhof's current law. For our purposes let us express this principle as follows:

The sum of unknown currents leaving a node is equal to the sum of known currents entering to the node.

The unknown currents leaving the node can be divided in three groups, which are mutually independent:

A. Currents leaving through resistances (As illustrated in Fig. 6.4 on page 90)

B. Currents which are dependent on voltages, but not through resistances (like those in voltage controlled current sources)

C. Currents which are independent of voltages across elements (such as those in voltage sources or at the output of an ideal operational amplifier)

For each current present in group C, we need another equation (element equation) or else a condition (like a potential being known).

An important remark for our goals is the fact that groups A y B will generate an expression of the form

$$Y_{k1} V_1 + Y_{k2} V_2 + Y_{k3} V_3 + \cdots + Y_{k(N-1)} V_{N-1} \qquad (6.1)$$

which does not depend on currents from group C, and can be written by inspection. Also, it is easy to program for automatic generation, as done later.

Traditionally, since no calculators were available, it was of outmost importance to have as few equations as possible. For that reason, methods to avoid including currents from group C were introduced, and are still being presented in textbooks. This is a good practice, since sometimes it becomes really necessary to work with a reduced number of equations. It is interesting to remark that this practice is still so important that analysis where group C is included is called *Modified Nodal Analysis*!

I present the nodal analysis method in two steps. First, I follow the traditional textbook approach of just working with resistances, known currents and currents dependent on voltages. The method is quite direct and easily programmable. I present a program example for it

To avoid the presence of voltage unrelated currents at this stage, any voltage source may be converted into current source, as explained in section 5.4. Another way to avoid these currents without using transformations is the "supernode" method[1] is applied. This method can also be written by inspection. I discuss it but I do not think necessary to discuss programming for it, though.

In the second stage, I include unknown currents not dependent on voltages, that is, the modified nodal analysis (MNA). This is avoided in textbooks which are oriented to hand analysis. Yet, we work here with the calculator! No need to avoid complexity! Besides, MNA will allow us to apply our tools faster and easier. MNA is also programmable, but I limit here to set the guidelines only.

6.2 Resistances and current sources only circuits

Take a circuit with $n + 1$ nodes, where one of the nodes is ground. The circuit contains only "known" independent current sources (which may be symbolic), resistances and voltage dependent current sources. Our objective in this section to write by inspection the equation at each node m different from ground in the form

$$Y_{m1} V_1 + Y_{m2} V_2 + \cdots + Y_{mm} V_m + \cdots + Y_{mn} V_n = In_m \qquad (6.2)$$

The left hand side of (6.2) represents, after algebraic manipulation, the sum of currents leaving the node. The right hand side the sum of currents entering to the node. In this equation:

- $V_1, V_2, \ldots$ are the *node potentials* for nodes 1, 2, ...
- Coefficients Y_{km} are called *nodal admittances* or *nodal conductances*
- I_m is called the *nodal current* at node m.

For convenience, the voltages and currents of the elements will be written with small letters, while potentials and node currents with capital letters. When this convention has not been followed, it must be clear from context what is the meaning of the variable.

---

[1]The correct name for this method should be "cut-set" method. However, "supernode" is already accepted by the community.

6.2. RESISTANCES AND CURRENT SOURCES ONLY CIRCUITS

The set of these nodal equations can be written in matrix form as

$$\mathbf{Yn\, Vn = In} \tag{6.3}$$

6.2.1 Theoretical principles

This subsection is included because not all textbooks have a section on how to write nodal equations by inspection. If you already know this principle, you may skip this part.

The basis for node equations is Kirchhoff's Current Law, which states that the sum of currents leaving a node is equal to the sum of currents entering the node. In Fig. 6.3, let us consider node m, with the following

> CONVENTION: currents leaving from the node are those through resistances and voltage-dependent current sources, and entering are known currents from independent sources.

It follows from this convention that

I_m is the sum of known currents entering node m: $I_m = \sum i_j$ \hfill (6.4)

Because of the convention, any known current leaving the node is taken with negative sign.

Let us now consider an isolated node m as shown in Fig. 6.3.

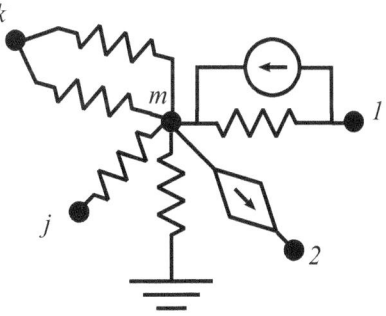

Figure 6.3: Isolated node m in circuit.

The nodal admittances Y_{mj} in (6.2) consists of two components: one, G_{mj} due to resistances and another one, go_{mj} arising from voltage-dependent current sources. That is,

$$Y_{mj} = G_{mj} + go_{mj} \tag{6.5}$$

Since each component is independent of the other, we deal with them separately.

Component of G_{mj} of Y_{mj} due to resistances.

Looking at Fig. 6.4, we can state the following fact for resistances (conductances):

For each resistance R connected to node m and node j, there will be a current leaving the node in the form

$$\frac{V_m - V_j}{R} = G(V_m - V_j) = GV_m - GV_j \qquad (6.6a)$$

And for each resistance connected between m and ground there is a current of the form

$$\frac{V_m}{R} = GV_m \qquad (6.6b)$$

Here $G = 1/R$ is the conductance of the resistance R.

Hence, we can conclude that when regrouping terms in the equation of node m, the coefficient of V_m will be the sum of all conductances connected to node m, while the coefficient of V_j for $j \neq m$ will be the negative of the sum of conductances connecting node m to node j, that is, the conductances shared by both nodes. Mathematically, this is expressed by

$$G_{mj} = \begin{cases} +\sum \text{Conductances connected to node } m & \text{for } j = m \\ -\sum \text{Conductances connected to both nodes } m \text{ and } j & \text{for } j \neq m \end{cases} \qquad (6.7)$$

Referring to the equation in matrix form, we can illustrate resistance at node m contributing to the row of the equation as illustrated in Fig. 6.4. Labels over the row show the columns and respective variables.

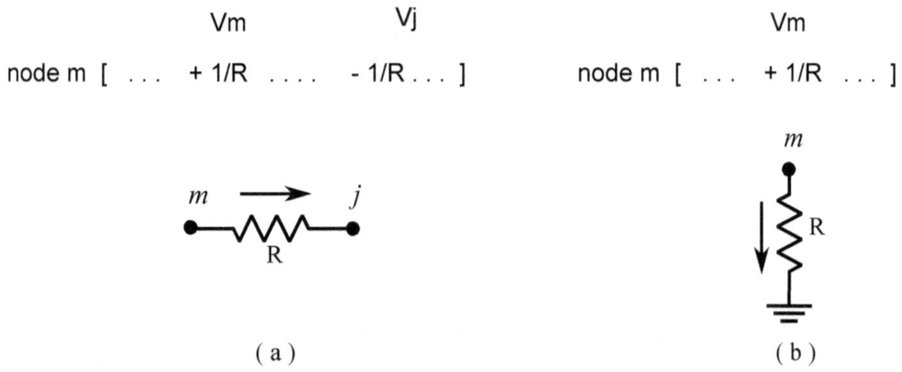

$$\text{node } m \begin{bmatrix} \cdots & \overset{V_m}{+1/R} & \cdots & \overset{V_j}{-1/R} & \cdots \end{bmatrix} \qquad \text{node } m \begin{bmatrix} \cdots & \overset{V_m}{+1/R} & \cdots \end{bmatrix}$$

(a)   (b)

Figure 6.4: Currents leaving node m through resistances.

6.2. RESISTANCES AND CURRENT SOURCES ONLY CIRCUITS

Voltage-controlled current sources

Unknown currents leaving from nodes include those through voltage controlled current sources of the type $go\,Vx = go(V_p - V_q)$. Since each source leaves from a node m and enters a node h, only coefficients in the equations at those two nodes are affected. Assuming connections as illustrated in Fig. 6.5, we can write the equations as

$$\text{Node } m: \overbrace{G_{m1}V_1 + \cdots + G_{mp}V_p + \cdots + G_{mq}V_q + \cdots}^{\text{Component of R's}} + g_o(V_p - V_q) = In_m$$
$$\text{Node } h: \underbrace{G_{h1}V_1 + \cdots + G_{hp}V_p + \cdots + G_{hq}V_q + \cdots}_{\text{Component of R's}} - g_o(V_p - V_q) = In_h$$

After some Algebra, we arrive at

$$\text{Node } m: G_{m1}V_1 + \cdots + (G_{mp} + g_o)V_p + \cdots + (G_{mq} - g_o)V_q + \cdots = In_m \quad (6.8a)$$

$$\text{Node } h: G_{h1}V_1 + \cdots + (G_{hp} - g_o)V_p + \cdots + (G_{hq} + g_o)V_q + \cdots = In_h \quad (6.8b)$$

```
                    Vp              Vq                              Vp              Vq
node m  [... Gmp + go   ... Gmq - go ...]    node m  [... Gmp + (a/R) ... Gmq - (a/R) .
node h  [... Ghp - go   ... Ghq + go ...]    node h  [... Ghp - (a/R) ... Ghq + (a/R) .
```

(a)   (b)

Figure 6.5: Voltage controlled current source and nodal coefficients.

Fig. 6.5 also illustrates how the source is inserted into the equations. Notice that the current controlled source was converted to voltage controlled source using Ohm's law. If you have problems remembering this scheme, then write down the equations first with the controlled sources separated from resistances and rearrange coefficients as explained above.

6.2.2 Rule to generate the equations

For convenience, let us repeat the rules to write by inspection the set of equations for the simplest case when the circuit consists of only independent current sources and resistances. The theoretical basis has been given in section 6.2.1 below.

At each node m, the nodal equation is written using the following rules:

$$I_m \text{ is the sum of known currents entering node } m: I_m = \sum i_j \qquad (6.9)$$

A current source entering the node is positive, and negative if it leaves the node.

$$G_{mj} = \begin{cases} +\sum \text{Conductances connected to node } m & \text{for } j = m \\ -\sum \text{Conductances connected to both nodes } m \text{ and } j & \text{for } j \neq m \end{cases} \qquad (6.10)$$

Two examples applying (6.4) and (6.10) will illustrate the algorithm. In both examples, equations are written in matrix form. The first one goes step by step, showing clearly the application of the criteria. The second one is straightforward, but the reader is encouraged to verify each matrix component.

Example 6.1 *A step by step example, Fist Part:*

Let us start with step-by-step setting of the nodal equations for the circuit in Fig. 6.6. Nodes are labeled 1, 2 and 3. We write the matrix $\mathbf{Y_n}$ and vector $\mathbf{I_n}$ by inspection using (6.10) and (6.9).

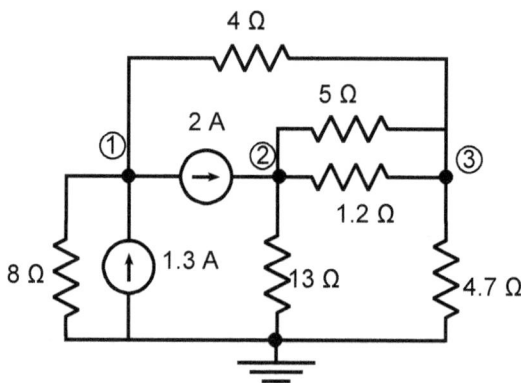

Figure 6.6: Circuit to analyze with nodal method.

Figures 6.7 and 6.8 show how the components of the matrices are found. The process is illustrated by taking subcircuits focused on the individual nodes 1, 2, and 3, keeping the components connected to them and seeing how those elements are shared with other nodes. Above each subcircuit, the row of "coefficients" in the equation of the respective node is displayed.

Consider node 1, at Fig. 6.7(a):

6.2. RESISTANCES AND CURRENT SOURCES ONLY CIRCUITS

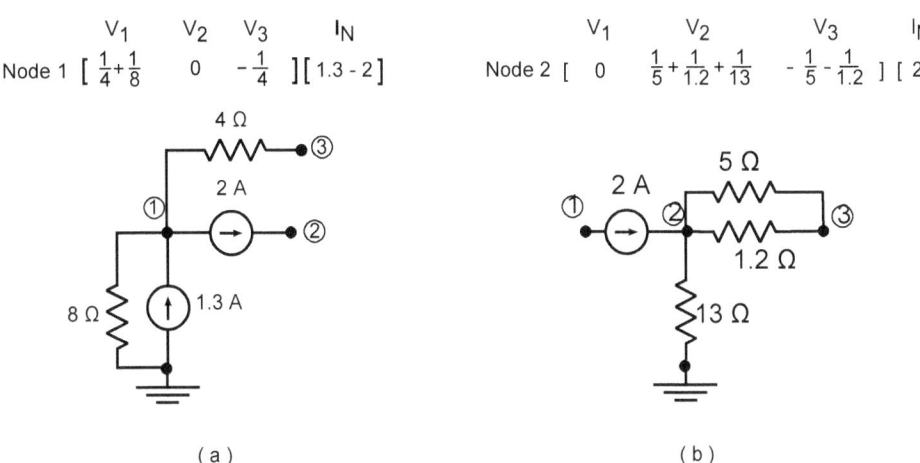

(a)   (b)

Figure 6.7: Forming matrix and equations for nodes (a) 1 and (b) 2.

1. Two resistances are connected to node 1: one of 8 Ω and another one of 4 Ω. Hence, the coefficient of V_1 in this equation is $(1/4 + 1/8)$; this is the element shown in the row vector of $\mathbf{Y_n}$ above the circuit, just under the label V1.

2. There are no resistances shared by nodes 1 and 2, so the coefficient of V_2 is 0, as shown under the label V2.

3. The resistance of 4 Ω is shared by nodes 1 and 3, so the coefficient of V_3 is -1/4.

4. For the respective row in the vector $\mathbf{I_n}$, we see a current source of 1.3 A entering into the node and another one of 2 A leaving the node. The element in the row of node 1 is therefore (1.3 - 2).

Look now at node 2, Fig. 6.7(b):

1. There are no resistances shared by nodes 1 and 2, so the coefficient of V_1 is 0, as shown under the label V1.

2. The three resistances connected to node 2 are of 5 Ω, 1.2 Ω, and 13 Ω. Hence, the coefficient of V_2 in this equation is $(1/5 + 1/1.2 + 1/13)$; this is the element shown in the row vector of $\mathbf{Y_n}$ above the circuit, just under the label V2.

3. The resistance of 4 Ω is shared by nodes 1 and 3, so the coefficient of V_3 is -1/4.

4. Finally, for the row in the vector $\mathbf{I_n}$, we see the current source of 2 A entering into the node. The element in the row of node 2 is therefore 2.

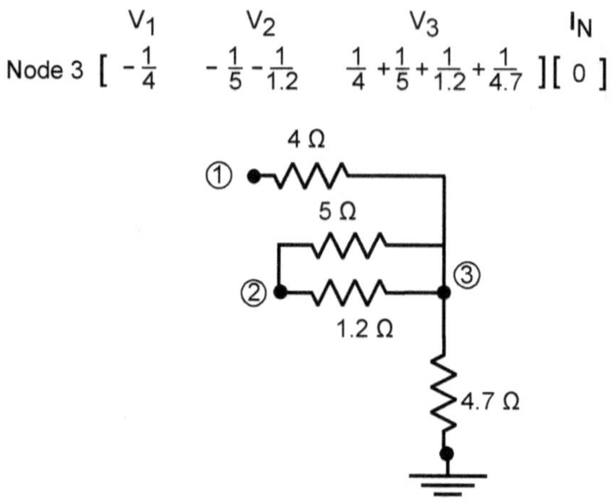

$$\text{Node 3} \begin{array}{cccc} V_1 & V_2 & V_3 & I_N \\ \left[-\frac{1}{4} \right. & -\frac{1}{5} - \frac{1}{1.2} & \left. \frac{1}{4} + \frac{1}{5} + \frac{1}{1.2} + \frac{1}{4.7} \right] & [\, 0 \,] \end{array}$$

Figure 6.8: Forming matrix and equations for node 3.

It should be clear that the coefficient of V_k in the equation of node m is the same as the coefficient of V_m in the equation of node k, since both are defined by the resistances shared by both nodes. This can be checked in the two steps above.

The reader is encouraged to find the equation for node 3. The set of equations can now be summarized in matrix form as

$$\begin{array}{c} \\ Node1 \\ Node2 \\ Node3 \end{array} \begin{array}{cccc} V_1 & V_2 & V_3 \\ \left[\begin{array}{ccc} \frac{1}{8} + \frac{1}{4} & 0 & -\frac{1}{4} \\ 0 & \frac{1}{5} + \frac{1}{1.2} + \frac{1}{13} & -\frac{1}{5} - \frac{1}{1.2} \\ -\frac{1}{4} & -\frac{1}{5} - \frac{1}{1.2} & \frac{1}{4} + \frac{1}{5} + \frac{1}{1.2} + \frac{1}{4.7} \end{array} \right] \end{array} \left[\begin{array}{c} V_1 \\ V_2 \\ V_3 \end{array} \right] = \begin{array}{c} In \\ \left[\begin{array}{c} 1.3 - 2 \\ 2 \\ 0 \end{array} \right] \end{array}$$
(6.11)

The solution for this set of equations is $[V_1 \, V_2 \, V_3]^T = [0.662, \, 5.332, \, 3.793]^T$

Let us now introduce this example onto the calculator, and solve.

Example 6.2 *A step by step example, Second Part:*

We introduce here the data for the previous example onto the calculator, illustrating the steps of Figs. 6.7 and 6.8. using the matrix/data editor. In your calculator, open the Data/Matrix Editor and create a 3x3 matrix **yn**.

Now enter the rows. First row:

1/4+1/8 ENTER 0 ENTER (-) 1/4 ENTER

The cursor should be now on the first element of the second row. Continue as

6.2. RESISTANCES AND CURRENT SOURCES ONLY CIRCUITS

0 ENTER 1/5 + 1/1.2 + 1/13 ENTER (-) 1/5 - 1/1.2 ENTER

and then to the third row

(-) 1/4 ENTER (-) 1/5 - 1/1.2 ENTER 1/5 + 1/1.2 + 1/4 + 1/4.7 ENTER

Go back to the home environment and check your matrix by calling it. With three decimal figures the calculator should display

$$\begin{bmatrix} .375 & 0.000 & -.250 \\ 0.000 & 1.110 & -1.033 \\ -.250 & -1.033 & 1.496 \end{bmatrix}$$

Next, create a new 3x1 matrix (vector) **In** *– call it* in, *and enter the source values of the nodal currents:*

1.3 -2 ENTER 2 ENTER 0 ENTER

Check your vector and then enter

$$yn^{-1} \;\boxed{\times}\; in \;\boxed{\text{ENTER}}\; \rightarrow \begin{bmatrix} .662 \\ 5.332 \\ 3.793 \end{bmatrix}$$

If you prefer to use the augmented matrix, the rows to enter are

First row:

1/4+1/8 ENTER 0 ENTER (-) 1/4 ENTER 1.3 -2 ENTER

Second row:

0 ENTER 1/5 + 1/1.2 + 1/13 ENTER (-) 1/5 - 1/1.2 ENTER 2 ENTER

And third row:

(-) 1/4 ENTER (-) 1/5 - 1/1.2 ENTER 1/5+1/1.2+1/4+1/4.7 ENTER 0 ENTER

Your matrix y *must now be displayed as*

$$\begin{bmatrix} .375 & 0.000 & -.250 & -.700 \\ 0.000 & 1.110 & -1.033 & 2.000 \\ -.250 & -1.033 & 1.496 & .000 \end{bmatrix}$$

Now apply the transformation rref(, and look for the solution at the last column:

$$\text{rref(y)}\ \boxed{\text{ENTER}} \rightarrow \begin{bmatrix} 1.000 & 0.000 & 0.000 & 0.662 \\ 0.000 & 1.000 & 0.000 & 5.332 \\ 0.000 & 0.000 & 1.000 & 3.793 \end{bmatrix}$$

The following example refers also to a circuit containing only resistances and independent current sources. Nodal equations are used again, but this time we write the nodal equations directly by inspection, following the previous rules, and use the results from the solution of the set of equations. If you have problems doing the first part, work the problem using an approach similar to example 6.1, where the individual sub circuits were isolated for easier identification.

Example 6.3 *For the circuit of Fig. 6.9, find Io and the power delivered by the current sources. Use nodal equations in the process.*

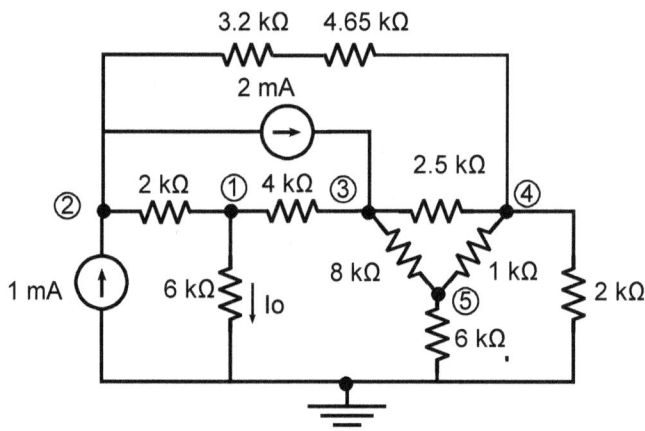

Figure 6.9: Another circuit to analyze with nodal method.

Solution: *Before proceeding to set up equations, first our strategy. Since our equations are for solving node potentials, we need to relate these potentials with the requested results.*

A generated power by a source is found by multiplying dc voltage and current, with the current leaving the + side of voltage. On the other hand, Io is the current in a resistance. Hence, the relationship between the node potentials and the requested results is as follows. For the generated powers:

generated by the 1 mA source: $P_1 = (1\,\text{mA})\,V_2$

generated by the 2 mA source: $P_2 = (2\,\text{mA})\,(V_3 - V_2)$

6.2. RESISTANCES AND CURRENT SOURCES ONLY CIRCUITS

For the current Io,

$$io = \frac{V_1}{6000}(A)$$

Now, there is no special request concerning the two resistances in series between nodes 2 and 4, so we can combine them in one equivalent resistance. Don't do it by yourself, just enter the expression in the formula.

That said, we are looking to solve for $\mathbf{Vn} = [V_1, V_2, \ldots, V_5]^T$ in

$$\mathbf{Yn\, Vn = In} \qquad (6.12)$$

where $\mathbf{Yn}$ is (with labels attached for easy reference)

	V_1	V_2	V_3	$\cdots$
$Node1$	$\frac{1}{6\,\text{E}3} + \frac{1}{4\,\text{E}3} + \frac{1}{2\,\text{E}3}$	$-\frac{1}{2\,\text{E}3}$	$-\frac{1}{4\,\text{E}3}$	$\cdots$
$Node2$	$-\frac{1}{2\,\text{E}3}$	$\frac{1}{2\,\text{E}3} + \frac{1}{3.2\,\text{E}3 + 4.65\,\text{E}3}$	0	$\cdots$
$Node3$	$-\frac{1}{4\,\text{E}3}$	0	$\frac{1}{8\,\text{E}3} + \frac{1}{2.5\,\text{E}3} + \frac{1}{4\,\text{E}3}$	$\cdots$
$Node4$	0	$-\frac{1}{3.2\,\text{E}3 + 4.65\,\text{E}3}$	$-\frac{1}{2.5\,\text{E}3}$	$\cdots$
$Node5$	0	0	$-\frac{1}{8\,\text{E}3}$	$\cdots$

	$\cdots$	$V4$	$V5$
$(Node1)$	$\cdots$	0	0
$(Node2)$	$\cdots$	$-\frac{1}{3.2\,\text{E}3 + 4.65\,\text{E}3}$	0
$(Node3)$	$\cdots$	$-\frac{1}{2.5\,\text{E}3}$	$-\frac{1}{8\,\text{E}3}$
$(Node4)$	$\cdots$	$\frac{1}{2.5\,\text{E}3} + \frac{1}{3.2\,\text{E}3 + 4.65\,\text{E}3} + \frac{1}{2\,\text{E}3} + \frac{1}{1\,\text{E}3}$	$-\frac{1}{1\,\text{E}3}$
$(Node5)$	$\cdots$	$-\frac{1}{1\,\text{E}3}$	$\frac{1}{1\,\text{E}3} + \frac{1}{6\,\text{E}3} + \frac{1}{8\,\text{E}3}$

and

$$\mathbf{In} = [0,\ 1\text{E}\boxed{(-)}3 - 2\text{E}\boxed{(-)}3,\ 2\text{E}\boxed{(-)}3,\ 0,\ 0]^T$$

In the above expressions I used "calculator notation E 3" to denote $\times 10^3$, and E-3 for $\times 10^{-3}$. Now enter the matrices in the calculator and store them to y and x, respectively, for quick keystroke. The key sequence for the solution is then, storing the solution to (z). To three decimal figures,

$$y^{-1} * x\ \boxed{\text{STO▶}}\ z \rightarrow \begin{bmatrix} 0.501 \\ -0.918 \\ 3.673 \\ 1.363 \\ 1.411 \end{bmatrix}$$

The elements in the vector **z** are, respectively, $V_1, V_2 \ldots V_5$. We can therefore find the desired answers for this problem using these vector components as follows:

For P_1 and P_2, respectively,

1 [EE] [(-)] 3 × z[2,1] [ENTER] → -917.806E⁻6

2 [EE] [(-)] 3 × (z[3,1] - z[2,1]) [ENTER] → 9.182E⁻3

Thus, the 1 mA source is actually absorbing 918 µW while the 2 mA source is generating 9.18 mW.

For current Io: z[1,1] / 6000 → 83.526E⁻6.
Hence, this current is 83.526 µA.

6.2.3 Scaling units

If both sides of a node equation are multiplied by 1000, the results are unaffected. This means that we could enter the resistances in kΩ units and the current sources in mA units, without changing the values or units of the voltages (V=RI). This brief remark is useful for us, both in time saving and in numerical rounding, since numbers are not as small.

With this remark, the current vector **In** in the previous example can be entered as

$$\mathbf{In} = [0, \quad 1-2, \quad 2, \quad 0, \quad 0]^T$$

and the matrix **Yn** as

$$\mathbf{Yn} =$$

$$\begin{bmatrix}
V_1 & V_2 & V_3 & V_4 & V_5 \\
\frac{1}{6}+\frac{1}{4}+\frac{1}{2} & -\frac{1}{2} & -\frac{1}{4} & 0 & 0 \\
-\frac{1}{2} & \frac{1}{2}+\frac{1}{3.2+4.65} & 0 & -\frac{1}{3.2+4.65} & 0 \\
-\frac{1}{4} & 0 & \frac{1}{8}+\frac{1}{2.5}+\frac{1}{4} & -\frac{1}{2.5} & -\frac{1}{8} \\
0 & -\frac{1}{3.2+4.65} & -\frac{1}{2.5} & \frac{1}{2.5}+\frac{1}{3.2+4.65}+\frac{1}{2}+\frac{1}{1} & -\frac{1}{1} \\
0 & 0 & -\frac{1}{8} & -\frac{1}{1} & \frac{1}{1}+\frac{1}{6}+\frac{1}{8}
\end{bmatrix}$$

Smaller matrices, easier and faster to enter. Notice that if a conductance value is already entered in the matrix, it should be in mS units.

6.2.4 An example with dependent sources

Let us work an example with dependent sources to illustrate dealing with them. We take advantage of the example to also illustrate again the advantages of scaling units.

Example 6.4 *For the circuit in Fig. 6.10, write down the matrices* **Yn** *and* **In** *of the nodal equation* **Yn Vn** = **In**. *Apply scaling by* 10^3, *i. e., resistance units in kilohms, currents in mA and conductance units in mS.*

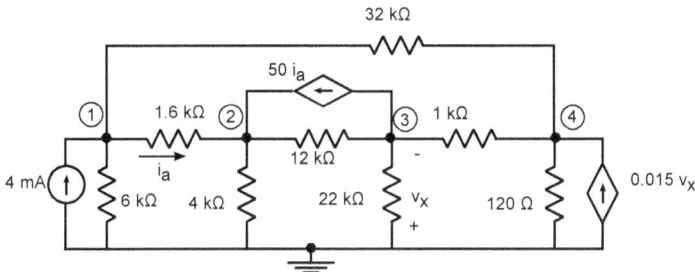

Figure 6.10: Voltage controlled current source and nodal coefficients.

Looking at the circuit, we see that there are two controlled current sources. Although the first one is current controlled, we convert it to voltage controlled using the fact that $i_a = (V_1 - V_2)/1.6k\Omega$. Hence, after applying scaling, this source adds

$$\frac{50}{1.6}(V_1 - V_2)$$

to the equation at node 3, and subtracts it at the equation of node 2, affecting coefficients Y_{13}, Y_{23}, Y_{12} and Y_{22}. Therefore

in equation of node 3: add $\frac{50}{1.6}$ to coefficient of V_1 and subtract $\frac{50}{1.6}$ from coefficient of V_2

in equation of node 2: subtract $\frac{50}{1.6}$ from coefficient of V_1 and add $\frac{50}{1.6}$ to coefficient of V_2

On the other hand, the voltage controlled current source $0.015\,v_x = -0.015 V_3$ will be subtracted at equation 4, affecting Y_{43} only, where we add 15 mS.

We can now work the matrix **Yn** directly as shown next, with labels attached for easy reference, and applying scaling,

$$\mathbf{Yn} = \begin{array}{c} Node1 \\ Node2 \\ Node3 \\ Node4 \end{array} \begin{bmatrix} \overset{V_1}{\frac{1}{6}+\frac{1}{1.6}+\frac{1}{32}} & \overset{V_2}{-\frac{1}{1.6}} & \overset{V_3}{0} & \overset{V_4}{-\frac{1}{32}} \\ -\frac{1}{1.6}-\frac{50}{1.6} & \frac{1}{1.6}+\frac{1}{4}+\frac{1}{12}+\frac{50}{1.6} & -\frac{1}{12} & 0 \\ 0+\frac{50}{1.6} & -\frac{1}{12}-\frac{50}{1.6} & \frac{1}{12}+\frac{1}{1}+\frac{1}{22} & -\frac{1}{1} \\ -\frac{1}{32} & 0 & -\frac{1}{1}+\mathbf{15} & \frac{1}{1}+\frac{1}{0.120}+\frac{1}{32} \end{bmatrix}$$
(6.13)

where bold entries are the ones due to the controlled sources.

Vector **In** is

$$\mathbf{In} = [4\ 0\ 0\ 0]^T$$

The result for this circuit is $[14.3, -2.8, -204.2, 305.3]^T$, whose values are the node potentials in Volt (V) units.

6.2.5 Modifying a circuit

Assume that after you have already set up the circuit equations, elements are added to, or deleted from, the circuit, without altering the number of nodes. That means that new elements are added connecting two nodes or a node to ground, in parallel to existing elements or else introducing one where there was none.

Figs. 6.4 and 6.5 illustrate how can we proceed without need to reintroduce matrices from the start. You can do it in the matrix editor, or directly on the command line. The rule is as follows:

Assume that you have already produced a matrix **Yn** for your circuit, and you connect a new resistance R between nodes k and j. Then you can obtain the matrix for the new circuit with

$$\text{Yn[k,k]} + 1/\text{R} \boxed{\text{STO}\blacktriangleright} \text{Yn[k,k]} \tag{6.14a}$$

$$\text{Yn[j,j]} + 1/\text{R} \boxed{\text{STO}\blacktriangleright} \text{Yn[j,j]} \tag{6.14b}$$

$$\text{Yn[k,j]} - 1/\text{R} \boxed{\text{STO}\blacktriangleright} \text{Yn[k,j]} \tag{6.14c}$$

$$\text{Yn[j,k]} - 1/\text{R} \boxed{\text{STO}\blacktriangleright} \text{Yn[j,k]} \tag{6.14d}$$

The above equations are applied whenever a new resistance is connected. However, we can work similarly other situations.

1. If the resistance is removed or disconnected, the signs for 1/R are changed in all equations (6.14)

6.2. RESISTANCES AND CURRENT SOURCES ONLY CIRCUITS

2. If the resistance is connected between node k and ground, apply only (6.14a)

3. If you find you have entered a wrong resistance R value, then "disconnect" the mistaken 1/R and "connect" the correct one in just one step in the equations.

These equations and remarks assume that no controlled source is affected with the modification. If there is any, study and change the corresponding entries.
The following exercise invites the reader to apply the above procedures. For this example, I recommend the reader to draw the modified schematics of the new circuit.

Example 6.5 *In example 6.4, the matrix $\mathbf{Y}n$ calculated in (6.13) is stored as Y in the calculator, while the source vector as Z. Displayed in engineering mode, Y is*

$$\begin{bmatrix} 822.92\text{E-}3 & -625.\text{E-}3 & 0.\text{E}0 & -31.25\text{E-}3 \\ 5.2083\text{E}0 & 32.208\text{E}0 & -83.333\text{E-}3 & 0\text{E}0 \\ 31.25\text{E}0 & -31.333\text{E}0 & 1.1288\text{E}0 & -1.\text{E}0 \\ -31.25\text{E-}3 & 0\text{E}0 & 14.\text{E}0 & 9.3646\text{E}0 \end{bmatrix}$$

Remember that all values are in mS units since resistances were entered in $k\Omega$ units and conductances in mS units!

Find the new $\mathbf{Y}n$ if the following modifications are made

1. *A new resistance of 3 $k\Omega$ is connected in parallel with the 1 $k\Omega$ resistance, between nodes 3 and 4, and the resistance in parallel with the independent source is erased.*

2. *The 120 Ω resistance value is changed to 12.3 $k\Omega$.*

To take into account the introduction of the 3 $k\Omega$ resistor:

Y[3,3] + 1/3 STO▶ Y[3,3]
Y[4,4] + 1/3 STO▶ Y[4,4]
Y[4,3] - 1/3 STO▶ Y[4,3]
Y[3,4] - 1/3 STO▶ Y[3,4]

To suppress the resistance of 6 $k\Omega$ in parallel with the independent source, between node 1 and ground:

Y[1,1] - 1/6 STO▶ Y[1,1]

Finally, to modify the 120 Ω resistance, taking into account the scaling:

Y[4,4] - 1/0.120 + 1/12.3 STO▶ Y[4,4]

6.2.6 Circuits with voltage sources using source transformation

For circuits containing voltage sources, either independent or voltage dependent, we can apply source transformation as well as the shifting theorem explained in section 5.4 so we arrive at a circuit containing only current sources, to which the procedure that has been explained can be applied. The transformations eliminate all the information for the current at the voltage source, and also reduces the number of nodes and as a consequence the number of equations. As a consequence, if you still need to calculate potentials or currents that have been eliminated, you must add additional equations. Working an example may be the best way to understand the method.

Example 6.6 *Let us work the circuit of example Fig. 6.11(a). When you apply source transformation to the independent source and the shifting theorem to the dependent source, followed by source transformations, you arrive at Fig. 6.11(b)[2].*

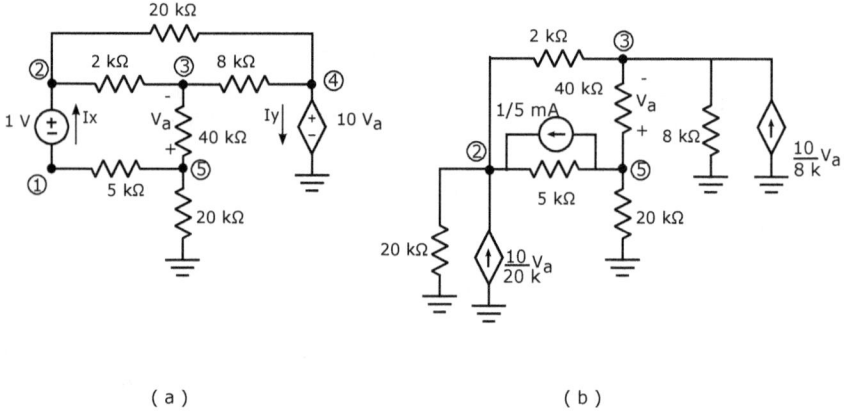

(a)   (b)

Figure 6.11: Analysis of circuit of Fig. 6.18 with current sources: (a) original circuit; (b) transformed circuit.

Observe that in the transformed circuit we have lost nodes 1 and 5, as well as currents I_x and I_y. All we need to solve now are three equations.

In the transformed circuit, using scaling for resistances in $k\Omega$ units, the equations are

$$\begin{array}{c} \\ Node2 \\ Node3 \\ Node5 \end{array} \begin{bmatrix} \overset{V_2}{\frac{1}{2}+\frac{1}{5}+\frac{1}{20}} & \overset{V_3}{-\frac{1}{2}+\frac{10}{20}} & \overset{V_5}{-\frac{1}{5}-\frac{10}{20}} \\ -\frac{1}{2} & \frac{1}{2}+\frac{1}{40}+\frac{1}{8}+\frac{10}{8} & -\frac{1}{40}-\frac{10}{8} \\ -\frac{1}{5} & -\frac{1}{40} & \frac{1}{40}+\frac{1}{20}+\frac{1}{5} \end{bmatrix} \begin{bmatrix} V_2 \\ V_3 \\ V_5 \end{bmatrix} = \begin{bmatrix} \overset{In}{\frac{1}{5}} \\ 0 \\ -\frac{1}{5} \end{bmatrix}$$

---
[2]Refer to section 5.4 if you need to refresh these concepts

6.3. OTHER CONSIDERATIONS FOR NODAL EQUATIONS

The result for this system is $[V_2 \quad V_3 \quad V_5]^{\mathrm{T}} = [-1.801 \quad -1.96 \quad -2.215]^{\mathrm{T}}$. *With*

$$V_2 - V_1 = 1 \quad and \quad V_4 = 10(V_5 - V_3)$$

we can get the remaining potentials. Similarly, equations at node 1 and 4 allow us to get the currents.

6.3 Other considerations for nodal equations

Before going on to other situations for node equations, let us introduce concepts and special applications. Even though introduced here, they can be applied to any later situation where known sources are not current sources.

6.3.1 Indefinite admittance matrix

When ground is not connected to any element of the circuit, the resultant matrix **Yn** is called *indefinite admittance matrix* (IAM). By extension, **In**, may be called indefinite nodal current vector. Both matrices **Yn** and **In** have the characteristic that the sum of the rows and columns are 0. As a consequence, the determinant of **Yn** is 0 an the set of equations does not have a solution. Yet, the IAM has many uses[3]. For this book we are interested in the following property.

When one of the nodes in the circuit is connected to ground, the corresponding column and row in the indefinite **Yn** *and* **In** *are deleted, and the matrix and vector become the definite nodal admittance matrix and vector that we already know.*

We will use this property in programming, but many other situations may be considered for application

6.3.2 Symbolic sources, time function and superposition

Using symbolic or sources, or sources of the form $Kf(t)$ as it was explained in Chapter 4, can be done using 1 as a value for the symbolic source, or K for the second one, and then multiplying by the symbolic variable or by $f(t)$. If there are more than one source we can apply superposition, using a matrix for the known side instead of a vector. This matrix statement also applies for superposition.

Let us illustrate the above remark with the circuit used for example 6.1.

Example 6.7 *For the circuit of Fig. 6.6 which is reproduced here as Fig. 6.12 for quick reference, find:*

1. *The contributions to the node potentials of the individual sources* 1.3 A *and* 2 A.

---

[3] For further information the reader may consult [Kiss68] [Moschytz1974].

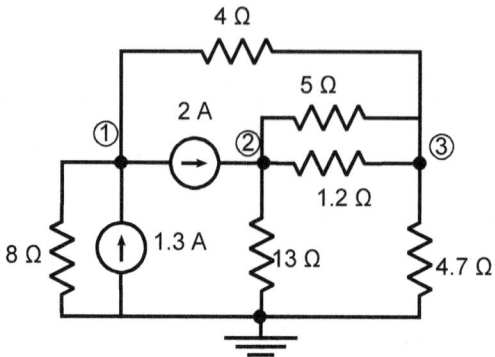

Figure 6.12: Circuit to analyze with nodal method.

2. The node potentials if the respective sources are $1.3\cos(200t)$ A and $2\sin(120t)$ A

3. The node potentials if these sources are instead IA and IB, respectively.

Solutions:

1) To find the individual contributions, we separate the terms that in the current vector **In** are of the form $1.3a + 2b$ and write individual columns for each source. Notice that the original elements can be interpreted as $(1)(1.3) + (-1)(2)$, $(0)(1.3) + (1)(2)$, and $(0)(1.3)+(0)(2)$. Now we write **In** as

$$\mathbf{In} = \begin{bmatrix} 1.3 & -2 \\ 0 & 2 \\ 0 & 0 \end{bmatrix}$$

Using now matrix $\mathbf{Y_n}$ generated before (see eq. 6.11), we have

$$\mathbf{Vn} = \mathbf{Yn}^{-1}\mathbf{In} = \begin{bmatrix} 5.048 & -4.478 \\ 2.137 & 3.730 \\ 2.372 & 1.283 \end{bmatrix}$$

The first column shows the nodal potentials due to the 1.3 A source, while the second column shows those generated by the 2 A source. If you also want the total, you can multiply this result by

$$\mathbf{z} = \begin{bmatrix} 1 \\ 1 \end{bmatrix}$$

2) For the sources $1.3\cos(200t)$ A and $2\sin(120t)$ A we can use the same matrices as before. However, now the result should be interpreted as

$$V_1 = 5.048\cos(200t) - 4.478\sin(120t)\,\mathrm{V}$$
$$V_2 = 2.137\cos(200t) + 3.730\sin(120t)\,\mathrm{V}$$
$$V_3 = 2.372\cos(200t) + 1.283\sin(120t)\ V$$

6.3. OTHER CONSIDERATIONS FOR NODAL EQUATIONS

This interpretation follows from the fact that now **z** is interpreted as

$$\mathbf{z} = \begin{bmatrix} \cos(200\,t) \\ \sin(200\,t) \end{bmatrix}$$

3) In this case, we use unit values for the sources, and simply interpret the results properly

$$\mathbf{In} = \begin{bmatrix} 1 & -1 \\ 0 & 1 \\ 0 & 0 \end{bmatrix}$$

and apply it to

$$\mathbf{Vn} = \mathbf{Yn}^{-1}\,\mathbf{In} = \begin{bmatrix} 3.883 & -2.239 \\ 1.644 & 1.865 \\ 1.825 & 0.6415 \end{bmatrix}$$

Therefore,

$$V_1 = 3.883\,\mathrm{IA} - 2.239\,\mathrm{IB}$$
$$V_2 = 1.644\,\mathrm{IA} + 1.865\,\mathrm{IB}$$
$$V_3 = 1.825\,\mathrm{IA} + 0.6415\,\mathrm{IB}$$

Observe that, because of the proportionality property, we could have got the coefficients from the previous item by dividing the first term coefficient by 1.3 and the second one by 2.

Again, the last interpretation follows because now

$$\mathbf{z} = \begin{bmatrix} \mathrm{IA} \\ \mathrm{IB} \end{bmatrix}$$

6.3.3 Programming nodal equations I

Since our calculator is programmable, let us take advantage to generate the matrices **Yn** and **In** and solve for the equations. The basic algorithm for circuits with characteristics mentioned so far, that is, with resistances, current sources and voltage controlled current sources, is described in the following steps using indefinite admittance matrix. **Ground is denoted as node "N+1" for easy row and column deletion.**

1. **Number of nodes** Specify number of nodes N, excluding ground. Denote ground as N+1.
2. **Initialization** Initialize matrix $\mathbf{Yn} = 0$ of order (N+1)x(N+1), and vector $\mathrm{IN} = 0$ of order (N+1)x1
3. **Resistance Subroutine** For each resistance of value R ($\neq 0, \infty$) connected between nodes j and k do:
    1. $Yn[j,j] = Yn[j,j] + 1/R$

2. $Yn[k,k] = Yn[j,j] + 1/R$
3. $Yn[j,k] = Yn[j,k] - 1/R$
4. $Yn[k,j] = Yn[k,j] - 1/R$

4. **Voltage Controlled Current Source Subroutine** If there are VCCS's, then, for each source with transconductance g going from node j to node k, and controlled by $(V_p - V_q)$ do:

    1. $Y[j,p] = Y[j,p] + g$
    2. $Y[j,q] = Y[j,q] - g$
    3. $Y[k,p] = Y[k,p] - g$
    4. $Y[k,q] = Y[k,q] + g$

5. **Current source Subroutine** For each current source of value I ($\neq 0, \infty$) going from node j to node k do:

    1. $In[j,1] = IN[j,1] - I$
    2. $In[k,1] = IN[k,1] + I$

6. **Create definite Yn:** Delete both row and column (N+1) of **Yn** and row (N+1) of **In**

7. **Solve for Vn:** If $|\mathbf{Yn}| \neq 0$ then $\mathbf{Vn} = \mathbf{Yn}^{-1}\mathbf{In}$

In simulators like SPICE, ground is denoted as 0. If you prefer to use such notation, then you should check for ground before each update of elements in the matrices.

Preparing the inputs

We could write our programs in a completely interactive way. I find this rather cumbersome and prefer to describe the circuit before calling the program. The circuit is described using matrices. The number of nodes is **n** excluding ground. Ground is denoted as **n+1** in data.

1. An $m \times 3$ matrix **R** for the m resistances in the circuit. If $m = 0$, then enter 0 for this parameter. Each row k consists of three items:

    - Element $(k, 1)$ is a non-zero and non-infinite resistance value,
    - Elements $(k, 2)$ and $(k, 3)$ are the two nodes to which it is connected.

2. A $p \times 3$ matrix **Is** for the p current sources in the circuit. If Is=0 , then **In** and **Vn** will be 0. Such situation is of interest when we are interested in **Yn** for other uses. Each row k of **Is** consists of three items:

    - Element is[k,1] is the finite current source value. It may be 0.
    - Element is[k,2] is the node from where the current is leaving,
    - Element is[k,3] the node to which it is entering.

6.3. OTHER CONSIDERATIONS FOR NODAL EQUATIONS

3. Input gs. This value is 0 if the circuit contains no VCCS. Otherwise, it is a $q \times 5$ matrix. In the k-th row,

- Element gs[k,1] is a finite transconductance g value.
- Element gs[k,2] is the node from where the current is leaving,
- Element gs[k,3] the node to which the current is entering.
- Element gs[k,4] is the node of the positive reference for the controlling voltage, and
- Element gs[k,5] is the node of the negative reference for the controlling voltage.

Fig. 6.13 illustrates how the rows are structured for each of the matrices.

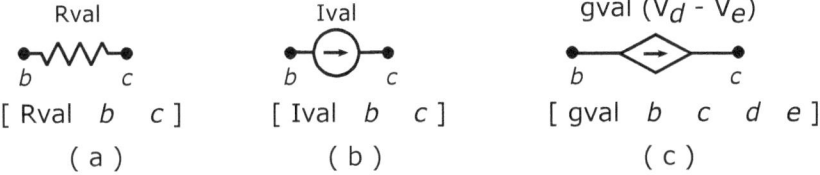

Figure 6.13: Inputs structure for program nodal1

Examples of descriptions: To illustrate the input circuit description, let us look at some previous examples.

A) For the circuit in Fig. 6.1 on page 92 of example 6.1:

$$n = 3 \quad r = \begin{bmatrix} 8 & 1 & 4 \\ 4 & 1 & 3 \\ 13 & 2 & 4 \\ 1.2 & 2 & 3 \\ 5 & 2 & 3 \\ 4.7 & 3 & 4 \end{bmatrix} \quad \text{is} = \begin{bmatrix} 1.3 & 4 & 1 \\ 2 & 1 & 2 \end{bmatrix} \quad \text{gs} = 0$$

B) For the circuit in Fig. 6.9 on page 96 of example 6.3:

$$n = 5 \quad r = \begin{bmatrix} 2000 & 1 & 2 \\ 4000 & 1 & 3 \\ 6000 & 1 & 6 \\ 3.2\text{E}3 + 4.65\text{E}3 & 2 & 4 \\ 2.5\text{E}3 & 4 & 3 \\ 8000 & 3 & 5 \\ 1\text{E}3 & 4 & 5 \\ 6000 & 5 & 6 \\ 2000 & 4 & 6 \end{bmatrix} \quad \text{is} = \begin{bmatrix} 1\text{E}^{-3} & 6 & 2 \\ 2\text{E}^{-3} & 2 & 3 \end{bmatrix} \quad \text{gs} = 0$$

Observe that in the first column we can specify an equivalent resistance showing the formula for the calculation. Also, I have mixed scientific and normal notation to make clear that you are not bound to one type of description.

C) Finally, for the circuit in Fig. 6.10 on page 99 of example 6.4:

$$n = 4 \qquad rs = \begin{bmatrix} 6\text{E}3 & 1 & 0 \\ 1.6\text{E}3 & 1 & 2 \\ 4\text{E}3 & 2 & 5 \\ 32000 & 1 & 4 \\ 12\text{E}3 & 2 & 3 \\ 22\text{E}3 & 3 & 5 \\ 1000 & 3 & 4 \\ 120 & 4 & 0 \end{bmatrix}$$

$$\text{is} = \begin{bmatrix} 4\text{E}^-3 & 5 & 1 \end{bmatrix} \qquad gs = \begin{bmatrix} 50/1.6\text{E}3 & 3 & 2 & 1 & 2 \\ 0.015 & 5 & 4 & 5 & 3 \end{bmatrix}$$

Programming the calculator

Fig. 6.14 on the facing page shows a program in the TI-89 environment following the given pseudo code. It takes as parameters the number of nodes n, excluding ground, and the matrices described in the previous section. In practice, the circuit might be void without any element, but I think you are not looking at this possibility.

The output of this program are the nodal admittance matrix yn, the nodal current vector in and the nodal potential vector vn. If there are no current sources, that is, is = 0, then in and vn are 0. The program can be easily modified (one line) if you also want the indefinite admittance matrix.

For easy understanding, the lines are commented on the right column; these comments are not part of the program but you may include them if you prefer to do so. Similarly, the lines are numbered for reference in explanation. Except for the case of checking if the circuit description is valid, I have not attempted any extra feature like data checking or other advisable programming precautions for a general case. I wrote the program originally for personal use and I assume that at least I will not provide wrong data.

The reader is encouraged to test the program for the different data. You should also be aware that the program presented here is just an example of a possible listing. The name of the program may be included in your custom menu so you can call it easily.

This program does not consider circuits with voltage sources or current dependent sources whose controlling current is not voltage dependent. However, as we will be able to see, the program can be used as a sub routine for a more general situation where these elements are considered.

6.4. CIRCUITS WITH VOLTAGE SOURCES AND OTHER ELEMENTS

```
1    :nodal1(n,r,is,gs)
2    :Pgrm
3    :Local a,b,c,d,e,t
4    :newMat(n+1,n+1)→ yn           Initializing the indefinite admmitance matrix
5    :For t,1,rowDim(r)             Start Yn construction with resistances
6    :r[t,1]→a::r[t,2]→b            For simpler notation in writing: a= value of R;
7    :r[t,3]→c                      R connected to nodes b,c
8    :yn[b,b]+1/a→yn[b,b]           Updating $Y_{bb}$
9    :yn[c,c]+1/a→yn[c,c]           Update $Y_{cc}$
10   :yn[b,c]-1/a→yn[b,c]           Updating $Y_{bc}$ and
11   :yn[c,b]-1/a→yn[c,b]           Updating $Y_{cb}$
12   :EndFor                        End Resistance subroutine
13   :If gs/=0 Then                 Continue Yn when there are VCCS's
14   :For t,1,rowDim(gs)            Start introducing dependent sources
15   :gs[t,1]→a::gs[t,2]→b          For simpler notation in writing: a=gm;
16   :gs[t,3]→c                     leaving from node b, entering node c
17   :gs[t,4]→d:  gs[t,5]→e         controlled by Vd-Ve
18   :yn[b,d]+a→yn[b,d]             Updating $Y_{bd}$
19   :yn[b,e]-a→yn[b,e]             Updating $Y_{be}$
20   :yn[c,d]-a→yn[b,d]             Updating $Y_{cd}$
21   :yn[c,e]+a→yn[b,e]             Updating $Y_{ce}$
22   :EndFor                        End reading gs
23   :EndIf                         Finish VCCS subroutine
24   :subMat(yn,1,1,n,n)→ yn        Definite admittance matrix
25   :                              Blank line for easy reading
26   :newMat(n+1,1)→ in             Initializing the indefinite nodal current vector
27   :If is/=0 Then                 If no current sources, only Yn is obtained
28   :For t,1,rowDim(is)            Start In construction with independent sources
29   :is[t,1]→a::is[t,2]→b          For simpler notation in writing: a=I;
30   :is[t,3]→c                     leaving from node b, entering node c
31   :in[b,1]-a→in[b,1]             Updating $In_b$
32   :in[c,1]+a→in[c,1]             Updating $In_c$
33   :EndFor                        End Independent currents subroutine
34   :EndIf                         End is/=0 case
35   :                              Blank line for easy reading
36   :subMat(in,1,1,n,1)→ in        Definite in
37   :If det(yn)/=0 Then            If circuit is valid
38   :yn^-1*in→ vn                  Solving for nodal voltage vector vn
39   :Else                          Otherwise
40   :Disp ''Description not valid''  Message for invalid description
41   :EndIf                         End vn calculation
42   :EndPrgm                       Exit Program
```

Figure 6.14: Example of a program for nodal analysis (resistances and current sources) using TI-89.

6.4 Circuits with voltage sources and other elements

As stated before, the method of nodal analysis is based on using node potentials as voltage variables and writing down the set of current equations for the nodes. Now, an actual connection of elements to a node m contains more variety than the one illustrated in Fig. 6.3 on page 89, as illustrated next in Fig. 6.15.

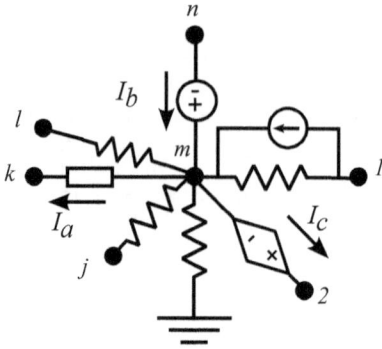

Figure 6.15: Voltage controlled current source and nodal coefficients.

These additional elements such as voltage sources, ideal operational amplifiers, and other type of elements, introduce unknown currents which are not voltage dependent. In the figure, these currents are exemplified by I_a, I_b, and I_c. Nevertheless, the current equation is still valid and may now be written as

$$\underbrace{Y_{m1} V_1 + \ldots + Y_{mn} V_n}_{\text{From resistances and VCCS}} + \underbrace{I_a - I_b + I_c}_{\text{non-voltage controlled}} = In_m \qquad (6.15)$$

Observe that the component from resistances and VCCS has already been presented in the previous section and can be written by inspection. The unknown currents are simply added to the left hand side (with + sign for currents leaving, - sign for currents entering the node).

Two approaches can be taken to deal with this type of equations. The first one, traditionally followed in textbooks, is to reduce the number of equations so as to eliminate the presence of these unknown currents. This can be done either by source transformation as explained in subsection 6.2.6, or else by working with the equations so that no current or only a selected subset of current appears. In my opinion (*and that's my opinion*), one mistake that textbooks make is to not inform the student that what is done is simply to skip some steps in the system equation solving process. I think this is a mistake because many students find it quite difficult then to solve for those currents when they are needed!

The second and general approach, is the improperly called Modified Nodal Analysis (MNA)[4], which consists in including all unknowns in the system. Let us start with this general method since the other ones are just a modification of it. Besides, we are using a good calculator! No need to fear having many equations.

---

[4]It receives this name because the first approach was historically the preferred one due to the number of equations that must be solved, among other reasons

6.5 Modified Nodal Analysis (MNA)

For the system of equations to have a unique solution, the number of unknowns must be equal to the number of equations, or we might reduce the number of unknowns with a known constraint upon a variable. In other words,

> For each unknown current present in the circuit, we must have either another equation or a constraint on the set of variables.

From this statement, we set of node equations by inspection for all the nodes, and then we add the additional equations or work the constraints. For the moment being, let us consider the cases illustrated in Fig. 6.16. The equations/constraints that are added or considered for the set are as follows:

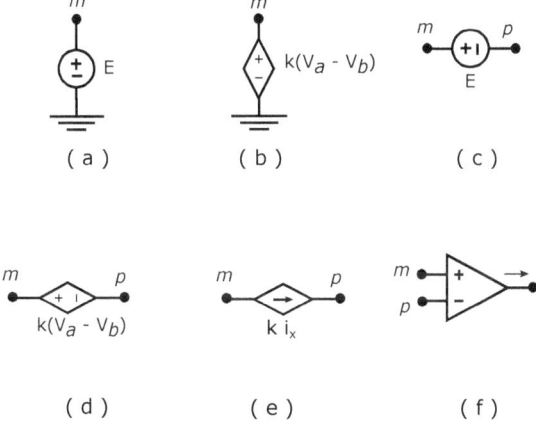

Figure 6.16: Cases for non voltage dependent currents: (a) Grounded source; (b) grounded VCVS; (c) One floating source; (d) One floating VCVS; (e) Grounded path of independent sources

(a) Grounded source: In this case we have the additional equation

$$V_m = E \qquad (6.16a)$$

An alternative is to include this value as a constraint directly on the set of equations and "pass" the variable to the other side in each case.

(b) Floating independent source: In this case the equation to be added is

$$V_m - V_p = E \qquad (6.16b)$$

(c) VCVS: In this case,

$$V_m - V_p - kV_a + kV_b = 0 \qquad (6.16c)$$

Naturally, one or two nodes may be ground, with potential of 0 V.

(d) **CCVS:** Assuming that the controlling current i_x is not a function of node potentials, we have

$$V_m - V_p - \alpha i_x = 0 \quad (6.16d)$$

(e) **CCCS:** Assuming that the controlling current i_x is not a function of node potentials, we do not need an additional variable here, since βi_x may be directly used in the equations of nodes m and p.

(f) **Ideal Operational Amplifier:** In this case,

$$V_m - V_p = 0 \quad (6.16e)$$

Let us work two examples. The first example is a DC circuit model for a bipolar transistor amplifier.

Example 6.8 *The circuit in Fig. 6.17 is the DC model of a transistor circuit, including parasitics ICE0 and Ro. For a nodal analysis of the circuit, the unknown currents that must be included are Ix and I_B. The current controlled source cannot be converted to a voltage controlled one. Hence, the total number of unknowns is 6, assuming that for the moment we do not convert V_1 to a known variable. For this example, let us take the following values:*

RC= 8 kΩ, R1 = 50 kΩ, R2 = 20 kΩ, RE = 4 kΩ, Ro = 80 kΩ, VCC = 28 V, VBE0= 0.65 V,
$\beta = 50$, and ICEO = 8 μA.

To simplify writing and keystrokes, resistance values are scaled, so all the currents will be in mA units (10 μA = .001 mA).

To clearly illustrate the procedure, let us start writing the individual equations at the nodes, and then the additional ones, the constrain equations, required to complete the number of necessary equations. The coefficients for the node potentials follow the same rules as before. For simplicity, I am omitting the terms with coefficient 0, although they should be considered in the calculator.

$$\text{Node 1: } \left(\frac{1}{8} + \frac{1}{50}\right) V_1 - \frac{1}{8} V_2 - \frac{1}{50} V_3 + I_x = 0$$

$$\text{Node 2: } -\frac{1}{8} V_1 + \left(\frac{1}{8} + \frac{1}{80}\right) V_2 - \frac{1}{80} V_4 + 50 I_B = -.008$$

$$\text{Node 3: } -\frac{1}{50} V_1 + \left(\frac{1}{50} + \frac{1}{20}\right) V_3 + I_B = 0$$

$$\text{Node 4: } -\frac{1}{50} V_2 + \left(\frac{1}{50} + \frac{1}{4}\right) V_2 - 51 I_B = .008$$

We have four equations with six variables. Now we add two equations introduced by the voltage sources:

6.5. MODIFIED NODAL ANALYSIS (MNA)

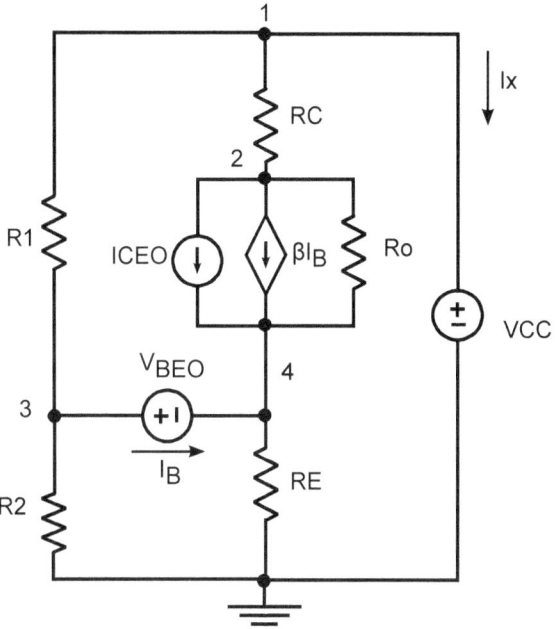

Figure 6.17: A first example for MNA.

$$V1 = 28$$

and

$$V_3 - V_4 = 0.65$$

The six equations may be incorporated in matrix form as $\mathbf{Y_{mn} X = I_{mn}}$, – "mn" is for "modified nodal" – as

$$\begin{bmatrix}
 & V_1 & V_2 & V_3 & V_4 & I_x & I_B \\
\text{Node 2} & \frac{1}{8}+\frac{1}{50} & -\frac{1}{8} & -\frac{1}{50} & 0 & 1 & 0 \\
\text{Node 3} & -\frac{1}{8} & \frac{1}{8}+\frac{1}{80} & 0 & -\frac{1}{80} & 0 & 50 \\
\text{Node 4} & -\frac{1}{50} & 0 & \frac{1}{50}+\frac{1}{20} & 0 & 0 & 1 \\
 & 0 & -\frac{1}{80} & 0 & \frac{1}{80}+\frac{1}{4} & 0 & -51 \\
\hline
28\text{ V src} & 1 & 0 & 0 & 0 & 0 & 0 \\
0.65\text{ V src} & 0 & 0 & 1 & -1 & 0 & 0
\end{bmatrix}
\begin{bmatrix} V_1 \\ V_2 \\ V_3 \\ V_4 \\ \hline I_x \\ I_B \end{bmatrix} =$$

$$\begin{array}{l}
\text{Node 1} \\
\text{Node 2} \\
\text{Node 3} \\
\text{Node 2} \\
\\
28\text{ V src} \\
0.65\text{ V src}
\end{array}
\begin{bmatrix} 0 \\ -0.008 \\ 0 \\ 0.008 \\ \hline 28 \\ 0.65 \end{bmatrix}$$

The vertical and horizontal separations have been introduced here to distinguish submatrices in a later explanation. They are not part of the equations, so don't try to enter them in the calculator!!.

After doing operations $\mathbf{Ymn}^{-1}\ \mathbf{Imn}$, showing the results with three decimal figures and, for convenience, in transposed form, we get

$$\mathbf{X}^T = \begin{bmatrix} V_1 & V_2 & V_3 & V_4 & I_x & I_B \\ 28 & 14.627 & 7.546 & 6.896 & -2.101 & 0.0318 \end{bmatrix}$$

From this result we read V_1= 28 V, V_2= 14.627 V, V_3= 7.546 V, V_4 = 6.896 V, I_x = -2.101 mA, and I_B = 0.032 mA = 31.8 μA.

Notice that in the partition made in the matrix $\mathbf{Ymn}$ in the previous example, the upper left submatrix of the coefficients as well as the upper subvector in the known side can be written by inspection following the rules already used in subsection 6.2.2. The other columns are really not difficult to write by inspection, but if you have problems with them, first write the second part of the equations separately or else try to visualize them.

Let us work another example.

Example 6.9 *Let us write now the MNA equations for the circuit of Fig. 6.18. This is the circuit used in example 6.6.*

Proceeding as we did before, the matrix $\mathbf{Ymn}$ *and vector* $\mathbf{Imn}$ *are given below. Again, notice the matrix subdivision shown here for later explanation. Resistance values are again in kΩ units so the currents are in mA units.*

6.5. MODIFIED NODAL ANALYSIS (MNA)

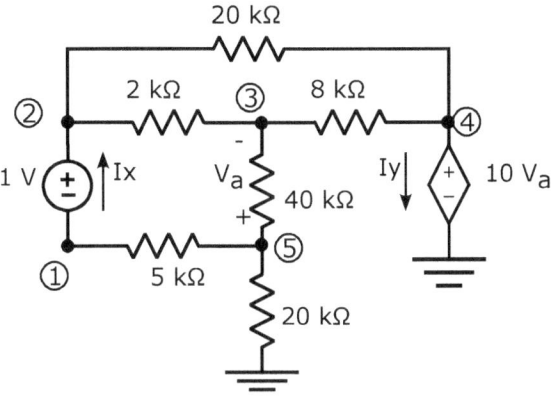

Figure 6.18: Another Example for Modified Nodal Analisis.

Ymn =

	V_1	V_2	V_3	V_4	V_5	I_x	I_y
N. 1	$\frac{1}{5}$	0	0	0	$-\frac{1}{5}$	1	0
N. 2	0	$\frac{1}{2}+\frac{1}{20}$	$-\frac{1}{2}$	$-\frac{1}{20}$	0	-1	0
N. 3	0	$-\frac{1}{2}$	$\frac{1}{40}+\frac{1}{2}+\frac{1}{8}$	$-\frac{1}{8}$	$-\frac{1}{40}$	0	0
N. 4	0	$-\frac{1}{20}$	$-\frac{1}{8}$	$\frac{1}{20}+\frac{1}{8}$	0	0	1
N. 5	$-\frac{1}{5}$	0	$-\frac{1}{40}$	0	$\frac{1}{40}+\frac{1}{5}+\frac{1}{20}$	0	0
1V	-1	1	0	0	0	0	0
VV	0	0	10	1	-10	0	0

$$\mathbf{Imn} = \begin{bmatrix} 0 \\ 0 \\ 0 \\ 0 \\ --- \\ 1 \\ 0 \end{bmatrix}$$

Creating these matrices in the calculator and entering

ymn ∧ (-) 1 × imn ENTER

The result, written in transposed notation for convenience, becomes

$$\mathbf{X}^T = \begin{matrix} V_1 & V_2 & V_3 & V_4 & V_5 & I_x & I_y \\ [-2.801 & -1.801 & -1.960 & -2.548 & -2.215 & 0.118 & -0.111] \end{matrix}$$

The first five entries are the respective node potentials, and the last two the currents in mA units.

When compared with example 6.6, we see that the values for potentials V_2, V_3, and V_5 are equal in both cases. Moreover, the two equations for the voltage sources mentioned in that example are included here. Finally, the equations for nodes 1 and 4 are also included. Thus, this method is more general and we could consider the one in example 6.6 as a system that results after algebraic work on the complete equations. I invite the reader to verify this statement.

Working MNA with known potentials

In example 6.8, we have an independent voltage source of 28 V which is grounded. This means that the potential $V_1 = 28$ V, as indicated in the constrain equation for the 28 V src. Although the way we proceeded is mathematically correct and also the way that it will result in the programs depicted in the next sections, it is quite annoying to write an additional equation for it, full of 0's in addition.

An alternate and most common procedure, is to take the potential $V_1 = 28$ already as a known value, eliminating the corresponding equation. This requires "to pass" the first column to the known side with a changed sign and including the value of the potential. The equations would look now as shown next:

$$\begin{matrix} & & V_2 & V_3 & V_4 & I_x & I_B & \\ \text{Node 1} & \begin{bmatrix} -\frac{1}{8} & -\frac{1}{50} & 0 & 1 & 0 \\ \frac{1}{8}+\frac{1}{80} & 0 & -\frac{1}{80} & 0 & 50 \\ 0 & \frac{1}{50}+\frac{1}{20} & 0 & 0 & 1 \\ -\frac{1}{80} & 0 & \frac{1}{80}+\frac{1}{4} & 0 & -51 \\ --- & --- & --- & --- & --- \\ 0 & 1 & -1 & 0 & 0 \end{bmatrix} \begin{bmatrix} V_2 \\ V_3 \\ V_4 \\ I_x \\ I_B \end{bmatrix} = \\ \text{Node 2} \\ \text{Node 3} \\ \text{Node 4} \\ --- \\ 0.65 \text{ V sc} \end{matrix}$$

$$\begin{matrix} \text{Node 1} \\ \text{Node 2} \\ \text{Node 3} \\ \text{Node 4} \\ --- \\ 0.65 \text{ V sc} \end{matrix} \begin{bmatrix} 0 - \left(\frac{1}{8}+\frac{1}{50}\right)(28) \\ -0.008 + \frac{28}{8} \\ 0 + \frac{28}{50} \\ 0.008 \\ --- \\ 0.65 \end{bmatrix}$$

6.5. MODIFIED NODAL ANALYSIS (MNA)

The reader can verify that, ignoring the first row, the equations obtained correspond to a shifting and source transformation of the VCC source. The complete set adds the equation for node 1 so the current Ix can be obtained.

An advantage of this procedure is that we can already visualize the actual known side.

Other cases of known potentials arise when there is a path of independent sources including ground, as illustrated in Fig. 6.19. Again, "passing" the known potentials to the right side is akin to successive applications of the shifting and conversion theorem, keeping the equations for the currents of the sources.

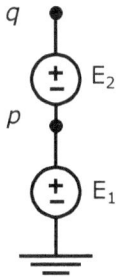

Figure 6.19: Known potentials arising from paths of independent sources. $V_p = E_1$, $V_q = E_1 + E_2$

6.5.1 Programming Modified Nodal Analysis

From examples 6.8 and 6.9, you can verify that **Ymn** and **Imn** may be partitioned as

$$\left[\begin{array}{c|c} \mathbf{Yn} & \mathbf{Ix} \\ \hline \mathbf{VEQ} & \mathbf{0} \end{array}\right] \quad \text{and} \quad \left[\begin{array}{c} \mathbf{In} \\ \hline \mathbf{VsVal} \end{array}\right] \quad (6.17)$$

respectively. Here, **Yn** and **In** are the nodal admittance matrix and nodal current vector already discussed in previous sections. The subroutines for these two submatrices can be found in § 6.3.3 on page 105, illustrated by the program in Fig. 6.14.

On the other hand, if there are M voltage sources with their respective unknown currents, this set generates submatrices **Ix** of order NxM, **VEQ** of order MxN, and the subvector **VsVal** of order M. N is the number of nodes excluding ground. These submatrices and vector arise from the M constraint equations added to the nodal equations. Finally, **0** is a 0-matrix of order MxM.

Remark: We will assume that current controlled current sources (CCCS) in which the current of control cannot be related to node potentials have the controlling current in a voltage source. This is illustrated in example 6.8 with current IB. Any other case can be worked by dealing with a manual edition of the matrices.

Since we already have subroutines for **Yn** and **In**, the pseudocode below works creating the matrices **Ymn** and **Imn** by matrix composition. To facilitate our task, we adopt the following

Convention: By definition, we assume that in the voltage sources the current is described by the passive convention of power, that is, it goes from the positive terminal to the negative terminal.

In other words, *for programming purposes, we will assume the current in the sources going from the + terminal to the negative terminal.* This means that if the current yields a negative value, the source is generating power and actual direction is leaving the positive terminal.

Circuit Description

Our pseudocode works with the description of the circuit using the number of nodes and a set of matrices as described below. The number of nodes N, as well as matrices **R,Is**, and **Gs** have been described before, but are repeated below for convenience. What we add for our analysis are the matrices that describe the way the voltage sources are connected. The set

Number of nodes The number of nodes N excluding ground

Matrix R for resistances For m resistances, an mx3 matrix, where each row describes a resistance connection using the same convention illustrated in Fig. 6.13. The matrix is 0 if there are no resistances.

Matrix Is for independent current sources For p independent current sources, a px3 matrix, where each row describes a source and its connections using the same convention illustrated in Fig. 6.13. The matrix is 0 if there are no independent current sources.

Matrix Gs for voltage dependent current sources For q VCCS, a qx5 matrix, where each row describes the dependent source and its connections using the same convention illustrated in Fig. 6.13. The matrix is 0 if there are no voltage controlled current sources.

Matrix Vs for independent voltage sources For s independent voltage sources, <u>whose currents do not control CCCS</u>, this is an sx3 matrix, where each row describes a source and its connections using the convention illustrated in Fig. 6.20. The matrix is 0 if there are no independent voltage sources.

Matrix Ks for voltage dependent voltage sources For t VCVS, a tx5 matrix, where each row describes a source and its connections using the convention illustrated in Fig. 6.20(b). The matrix is 0 if there are no dependent voltage sources.

6.5. MODIFIED NODAL ANALYSIS (MNA)

Matrix CCs for current dependent current sources For k CCCS, each one being controlled by a current in a voltage source, define a a kx6 matrix, where each row describes a voltage source and its connections, and the CCC source, using the convention illustrated in Fig. 6.20(c). The matrix is 0 if there are no dependent current controlled sources.[5]

We can work other situations by hand editing the matrices. Let us now illustrate the circuit description with examples.

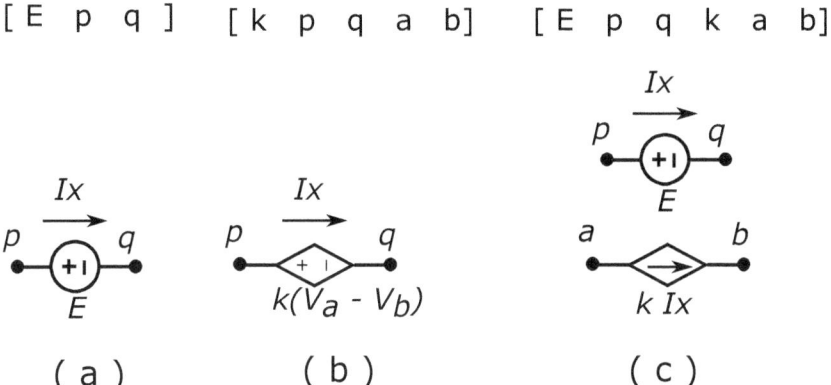

Figure 6.20: Matrix description of voltage sources for Modified Nodal Analisis.

Example 6.10 *The circuit of example 6.8, shown in 6.17 on page 113 is described by the following matrices:*

$$\mathbf{R} = \begin{bmatrix} 50E3 & 1 & 3 \\ 20E3 & 3 & 5 \\ 4E3 & 4 & 5 \\ 8E3 & 1 & 2 \end{bmatrix} \quad \mathbf{Is} = [\ 8E\boxed{(-)}6 \ \ 2 \ \ 4\ ] \quad \mathbf{Gs} = 0$$

$$\mathbf{Vs} = [\ 28 \ \ 1 \ \ 5\ ] \quad \mathbf{Ks} = 0 \quad \mathbf{CCs} = [\ 0.65 \ \ 3 \ \ 4 \ \ 50 \ \ 2 \ \ 4\ ]$$

On the other hand, the circuit of example ??, shown in 6.18 on page 115 is described by the following matrices:

$$\mathbf{R} = \begin{bmatrix} 5E3 & 1 & 5 \\ 2E3 & 2 & 3 \\ 20E3 & 2 & 4 \\ 8E3 & 3 & 4 \\ 40e3 & 3 & 5 \\ 20e3 & 5 & 6 \end{bmatrix} \quad \mathbf{Is} = 0 \quad \mathbf{Gs} = 0$$

---

[5] We assume a non-zero voltage source here. SPICE, the popular simulater, uses 0 V sources for these controlled sources

$$\mathbf{Vs} = [\ 1\ \ 2\ \ 1\ ]\quad \mathbf{Ks} = [\ 10\ \ 4\ \ 6\ \ 5\ \ 3\ ]\quad \mathbf{CCs} = 0$$

Program pseudocode for mna(n,r,is,gs,vs,ks,ccs)

One pseudocode, assuming the inputs described before, is the following:

Step 1 Generate submatrices **Yn** and **In** calling the program nodal1(n,r,is,gs) as subroutine.

Check for need of MNA analysis:

Step 2 IF (**Vs**=0 AND **Ks**=0 AND **CCS**=0) THEN
    Display message "No voltage sources"
    STOP
    ENDIF

Initialize submatrices and constraint index:

Step 3 **VEQ** = 0 of order Mx(N+1), **Ix** = 0 of order (N+1)x M and vector **Vsval** = 0 of order Mx1

Step 4 SourceIndex=0 for indexing purposes

Enter independent voltage sources:

Step 6 IF **Vs** $\neq$ 0 THEN
    FOR $h = 1$ to rowdimension(**Vs**)
      SourceIndex = SourceIndex+1
      E= **Vs**[h,1], p= **Vs**[h,2], and q= **Vs**[h,3]
      VEQ[SourceIndex,p] = 1
      VEQ[SourceIndex,q] = -1
      $\mathbf{VsVal}[SourceIndex, 1] = E$
      Ix[m,SourceIndex] = **Ix**[m,SourceIndex] + 1
      Ix[p,SourceIndex] = **Ix**[p,SourceIndex] -1
    ENDFOR
    ENDIF

Enter voltage dependent voltage sources:

Step 7 IF **Ks** $\neq$ 0 THEN
    FOR h = 1 to rowdimension(**Ks**)
      SourceIndex = SourceIndex+1   K= VS[h,1], p= **Ks**[h,2], and q= **Ks**[h,3]
      a= **Ks**[h,4] and b= **Ks**[h,5]

6.5. MODIFIED NODAL ANALYSIS (MNA)

          VEQ[SourceIndex,p] = **VEQ**[SourceIndex,p] + 1
          VEQ[SourceIndex,q] = **VEQ**[SourceIndex,q]-1
          VEQ[SourceIndex,a] = **VEQ**[SourceIndex,a] - k
          VEQ[SourceIndex,b] = **VEQ**[SourceIndex,b] + k
          Ix[p,SourceIndex] = **Ix**[p,SourceIndex] + 1
          Ix[q,SourceIndex] = **Ix**[q,SourceIndex] -1
      ENDFOR
    ENDIF

Enter current dependent current sources:

Step 8 IF **CCs** $\neq$ 0 THEN
      FOR h = 1 to rowdimension(**CCs**)
        SourceIndex = SourceIndex+1
        E= **CCs**[h,1], m= **CCs**[h,2], p = **CCs**[h,3]
        K= **CCs**[h,4], a= **CCs**[h,5], b = **CCs**[h,6]
        VEQ[SourceIndex,m] = 1
        VEQ[SourceIndex,p] = -1
        VsVal[SourceIndex,1] = E
        Ix[m,SourceIndex] = **Ix**[m,SourceIndex] + 1
        Ix[p,SourceIndex] = **Ix**[p,SourceIndex] -1
        Ix[n,SourceIndex] = **Ix**[n,SourceIndex] + K
        Ix[q,SourceIndex] = **Ix**[q,SourceIndex] - K
      ENDFOR
    ENDIF

Adjust dimensions of submatrices:

Step 9 Delete (N+1)-th column of **VEQ**, and row (N+1) of **Ix**

Compose Ymn and Imn and solve:

Step 10 Compose **Ymn** and **Imn** of expression (6.17) – See subsection §3.4.7 on page 34 –

    11.1 p= Augment(Yn,Ix)

    11.2 m= Augment(VEQ,0)

    11.3 Ymn = Augment(p;m)

    11.4 Imn = Augment(In;VsVal)

Step 11 IF det(**Ymn**)$\neq$ 0 THEN
        Vmn = **Ymn**$^{-1}$ **Imn**
    ELSE

Display "Matrix **Ymn** ill defined"
ENDIF

Exercise As an exercise for the reader, I suggest to use the above pseudocode and test it with the examples already presented.

Developing the program for mna

Since the program developed with the above pseudocode is rather large, for the sake of easy reading and printing, let me cut it in parts. I know that the pseudocode and program have some limitations. For example, the current dependent current sources have limitations. It does not include current controlled voltage sources whose controlling current is not voltage dependent. Other observations can be made. But, remember, this is an exercise in learning. And besides, the pseudocode and program are still very useful for a lot, if not most, of circuits you may find along the way. In any case, you can always edit by hand the matrices.

Fig. 6.21 shows the first part of the program. It covers the generation of **Yn** and **In**, as well as the initialization of submatrices **VEQ**, **Ix** and **VsVal**. Notice that this initialization requires to determine also the size of these submatrices. That is the number of rows for **VEQ** and **VsVal**, which is the same as the number of columns for **Ix**. This number is equal to (number of independent voltage sources + number of voltage controlled voltage sources + number of current controlled current sources).

```
1   :mna1(n,r,is,gs,vs,ks,ccs)
2   :Pgrm
3   :Local a,b,p,q,e,k,h,sourceindex
4   :nodal1(n,r,is,gs)                   Generate Yn and In
5   :If (vs=0 and ks=0 and ccs=0) Then   If no need of MNA
6   :Disp ''No voltage sources'' :stop   leave program
7   :Endif
8   :0 → h                               Determine size of submatrices
9   :If vs/=0 Then
10  :h+ rowDim(vs) → h                   Number of
11  :Endif                               independent V sources
12  :If ks/=0 Then                       plus
13  :h+ rowDim(ks) → h                   Number of
14  :Endif                               VCV sources
15  :If ccs/=0 Then                      plus
16  :  h+ rowDim(ccs) → h                Number of
17  :Endif                               CCS sources
18  :newMat(h,n+1)→ veq                  Initialize VEQ
19  :newMat(n+1,h)→ ix                   Initialize Ix
20  :newMat(h,1)→ vsval                  Initialize VsVal
```

Figure 6.21: Example of a modified nodal analysis program using the given pseudocode.

You can notice that lines 5, 6, and 7 are really unnecessary, since I imagine that you would not use this program unless you need it. But I introduced it anyway just for the sake of completeness.

6.5. MODIFIED NODAL ANALYSIS (MNA)

The second part of the program shown in Fig. 6.22 introduces the sources. It is in this section that you may add lines to include those devices that are not being considered. Namely, the ideal operational amplifiers and the current controlled voltage sources.

```
21   0 → sourceindex                              Init. Index
22  :If vs/=0 Then                                Introduce Ind. Sources
23  :For h,1,rowDim(vs)                           Start
24  :sourceindex + 1 → sourceindex                update index
25  :vs[h,1]→ E                                   source value E;
26  :vs[h,2] → p :vs[h,3]→ q                      terminals +p -q
27  :+1→ veq[sourceindex,p]                       Updating Veq
28  :-1→ veq[sourceindex,q]
29  :E→ vsval[sourceindex,1]                      Update VsVal
30  :ix[p,sourceindex]+1→ ix[p,sourceindex]       Updating Ix
31  :ix[q,sourceindex]-1→ ix[q,sourceindex]
32  :EndFor :Endif                                End independent sources.
33  :If ks/=0 Then                                Start dependent VCVS
34  :For h,1,rowDim(vs)                           Start
35  :sourceindex + 1 → sourceindex                update index
36  :vs[h,1]→ K                                   Gain K;
37  :vs[h,2] → p:  vs[h,3]→ q                     terminals +p -q
38  :ks[h,4]→a:  ks[h,5]→b                        controlled by Va-Vb
39  :veq[sourceindex,p]+1→ veq[sourceindex,p]     Updating Veq
40  :veq[sourceindex,q]-1→ veq[sourceindex,q]
41  :veq[sourceindex,a]-K→ veq[sourceindex,a]
42  :veq[sourceindex,b]+K→ veq[sourceindex,b]
43  :ix[p,sourceindex]+1→ ix[p,sourceindex]       Updating Ix
44  :ix[q,sourceindex]-1→ ix[q,sourceindex]
45  :EndFor :Endif                                End VCV sources.
46  :If ccs/=0 Then                               Introduce CCS sources
47  :For h,1,rowDim(ccs)                          Start
48  :sourceindex + 1 → sourceindex                update index
49  :vs[h,1]→ E                                   controlling source value E;
50  :vs[h,2] → p :vs[h,3]→ q                      terminals +p -q
51  :+1→ veq[sourceindex,p]                       Updating Veq
52  :-1→ veq[sourceindex,q]
53  :E→ vsval[sourceindex,1]                      Update VsVal
54  :ix[p,sourceindex]+1→ ix[p,sourceindex]       Updating Ix
55  :ix[q,sourceindex]-1→ ix[q,sourceindex]
56  :ccs[h,4]→ K                                  Gain K of CCS;
57  :ccs[h,5]→a :ccs[h,6]→b                       from a to b
58  :ix[a,sourceindex]+K→ ix[a,sourceindex]       Updating Ix
59  :ix[b,sourceindex]-K→ ix[b,sourceindex]
60  :EndFor :Endif                                End CCS sources.
```

Figure 6.22: Example of a modified nodal analysis program using the given pseudocode.

The (local) variable `sourceindex` is used to signal the row for the constraint equation and the column for the unknown current within the respective matrices. Notice that they are placed in the same order in which you have described them in the data matrices. If you desire to include more devices, you may continue here, after line 60 and before the first line (61) of the third part of this code, shown in Fig. 6.23. This last part of the code composes **Ymn** and **Imn**, and solves the equations.

```
61   :subMat(veq,1,1,sourceindex,n)→ veq              adjust veq
62   :subMat(ix,1,1,n,sourceindex)→ ix                Definite ix
63   :augment(yn,Ix) → p                              Upper subpartition of Ymn
64   :augment(veq,newMat(sourceindex,sourceindex))→ q Lower subpartition
65   :augment(p;q) → Ymn                              Compose Ymn
66   :augment(in;vsval) → Imn                         Compose Imn
67   :If det(Ymn)/=0 Then                             If circuit is valid
68   :Ymn∧ (-) 1 *Imn→ Vn                             Solve equations
39   :Else                                            Otherwise
40   :Disp ''Description not valid''                  Message for invalid description
41   :EndIf                                           End vn calculation
42   :EndPrgm                                         Exit Program
```

Figure 6.23: Example of a modified nodal analysis program using the given pseudocode.

Well, if you have created the above program, do not forget to include it in you custom menu and also to test it. Remember though that the most important step in the process is to interpret results. In particular, your solution **Vn** includes both node potentials and voltage source currents.

6.6 "Reducing" number of equations

Most textbooks, at least the introductory ones, avoid modified nodal analysis. Again, the reason is simple: they are oriented to hand analysis, so the student should focus in setting up systems with as few equations and variables as possible. This may be achieved by working directly on the equations, so as to reduce variables. But textbooks' authors usually prefer to work with equations directly. Two situations are considered here. The first one is by completely eliminating unknown currents. The second one consists in leaving one ore more currents in the set.

6.6.1 Eliminating all unknown currents by inspection: Supernodes

The method introduces the concept of "supernodes", a term that I also use just because it is well established, but don't like[6]. Fig. 6.24 illustrates this concept, where the nodes within the closed dashed line constitute the supernode.

In this figure, I have shown the equation that is assigned to the supernode, *assuming no resistances shared by the nodes within the closed curve*. If this is not the case, substitute Ypp and Yqq by Ypp + Ypq, and Yqq + Yqp.

---

[6] Actually, it is a concept known to circuit and graph theoreticians as "cut-set". But someone in the academic community introduced the term supernode, which has gained its place in basic circuit analysis textbooks.

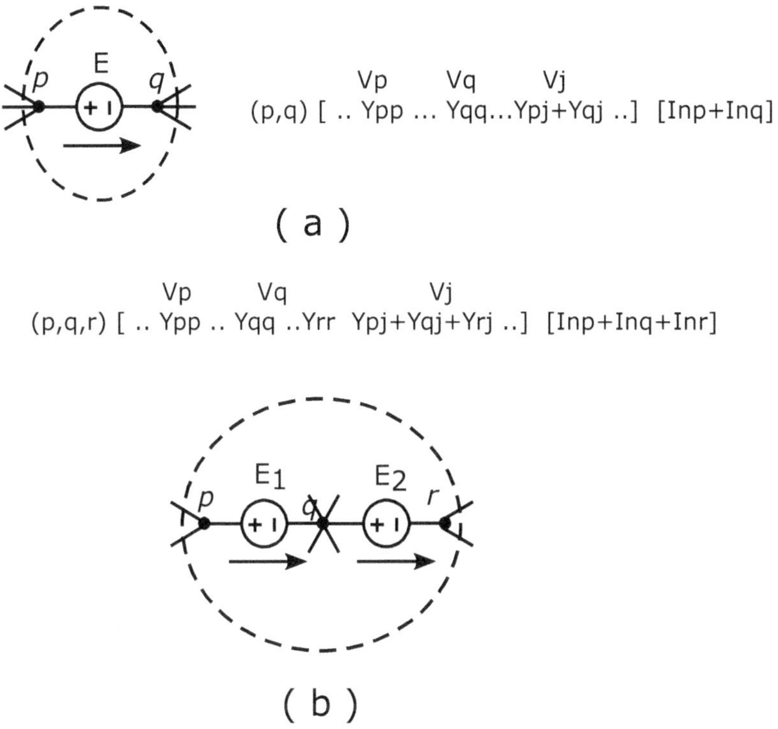

6.6. "REDUCING" NUMBER OF EQUATIONS

(a)

$$(p,q) \; [\; .. \; Y_{pp} \; ... \; Y_{qq} ... Y_{pj}+Y_{qj} \; ..] \; [I_{np}+I_{nq}]$$
with columns V_p, V_q, V_j.

(b)

$$(p,q,r) \; [\; .. \; Y_{pp} \; .. \; Y_{qq} \; .. Y_{rr} \; Y_{pj}+Y_{qj}+Y_{rj} \; ..] \; [I_{np}+I_{nq}+I_{nr}]$$
with columns V_p, V_q, V_j.

Figure 6.24: Supernode concept and the respective equations. If one node is ground, no equation is taken.

The method consists in not taking nodal equations at the nodes p, q, r,... . **Only** use the equation for the supernode together with the constraint equations arising from the voltage sources. *Notice that you cannot use supernodes if the current in the voltage source controls a dependent source!* This remark is illustrated by example 6.8, where the unknown current I_x controls a dependent source.

The reader can demonstrate that, mathematically, the equation of the supernode is given by the addition of the node equations for each of the nodes inside the closed curve. On the other hand, if one of the nodes is ground, we can eliminate the equation because it is the only with the unknown current, and thus is not necessary to solve for the rest of the variables.

Let us work examples to show how you can write the equations directly with the knowledge you have gained, saving time and effort while introducing equations in your calculator. Remember, try more exercises from your textbook of from other circuit books.

Example 6.11 *Let us start with the circuit shown in Fig. 6.18 on page 115 from example 6.9. Here, node 4 belongs to a grounded voltage source, so no equation is taken at this node. On the other hand, nodes 1 and 2 constitute a supernode created* with the 1 V source. With this information, we may write the matrix equation as shown below. Notice that there are five equations, for the five potentials. For the supernode, apply the principles shown in Fig. 6.24.

$$\begin{array}{c} \text{SN. (1,2)} \\ \text{N. 3} \\ \text{N. 5} \\ 1V \\ VV \end{array} \begin{bmatrix} \frac{1}{5} & \frac{1}{2}+\frac{1}{20} & \frac{1}{2} & -\frac{1}{20} & -\frac{1}{5} \\ 0 & -\frac{1}{2} & \frac{1}{40}+\frac{1}{2}+\frac{1}{8} & -\frac{1}{8} & -\frac{1}{40} \\ -\frac{1}{5} & 0 & -\frac{1}{40} & 0 & \frac{1}{40}+\frac{1}{5}+\frac{1}{20} \\ -1 & 1 & 0 & 0 & 0 \\ 0 & 0 & 10 & 1 & -10 \end{bmatrix} \begin{bmatrix} V_1 \\ V_2 \\ V_3 \\ V_4 \\ V_5 \end{bmatrix} = \begin{bmatrix} 0 \\ 0 \\ 0 \\ 1 \\ 0 \end{bmatrix}$$

I encourage the reader to solve this matrix equation and verify the the node potentials are the same as before.

Let us next work another example, this time with a known potential. For the known potential, "we pass it" to the known side, as always.

Example 6.12 *This example contains two grounded sources, one is known and the other is a dependent source. The circuit is shown in Fig. 6.25. In this circuit, I have shown explicitly the currents in the voltage sources. The current in the independent source will appear exclusively in the equation of node 4. Hence we can eliminate this equation. Being the last one in the set, this is easily done by eliminating the last row of the matrices. On the other hand, the equation of node 3 is subsituted by the equation of the VCVS. This can also be easily done.*

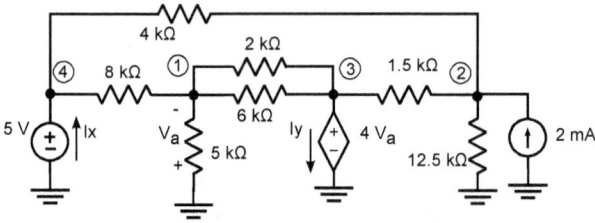

Figure 6.25: Example: circuit with two grounded voltage sources

The voltage controlled voltage source $4v_x$ is characterized by the constraint $V_3 = -4V_1$ or, taking into consideration that both potentials V_1 and V_3 are unknowns, by

$$4V_1 + V_3 = 0$$

. This equation is to be included in the set, instead of a nodal equation at node 3. Let us now write the equations **by inspection** considering the following facts:

- The unknown potentials are V_1, V_2 and V_3. Hence the coefficient matrix has three columns

6.6. "REDUCING" NUMBER OF EQUATIONS

- There are equations for nodes 1 and 2, but not for node 3. These equations follow the normal rules explained before.

- Instead of an equation for node 3, we use the equation for the controlled voltage source

The equation uses kΩ units for resistances, and mA units for current sources.

$$\begin{array}{c} Node1 \\ Node2 \\ VCVS \end{array} \begin{bmatrix} V_1 & V_2 & V_3 \\ \frac{1}{8}+\frac{1}{5}+\frac{1}{2}+\frac{1}{6} & 0 & -\frac{1}{2}-\frac{1}{6} \\ 0 & \frac{1}{1.5}+\frac{1}{12.5}+\frac{1}{4} & -\frac{1}{1.5} \\ 4 & 0 & 1 \end{bmatrix} \begin{bmatrix} V_1 \\ V_2 \\ V_3 \end{bmatrix} = \qquad (6.18)$$

$$\begin{array}{c} Node1 \\ Node2 \\ VCVS \end{array} \begin{bmatrix} \text{``}I_n - 5\,Y_{k4}\text{''} \\ \frac{5}{8} \\ 2+\frac{5}{4} \\ 0 \end{bmatrix}$$

I suggest the reader to write the MNA equations and then apply the procedures necessary to reduce the system to the one obtained in the examples. Let us work another example.

Applying superposition

As you might imagine, we can also apply superposition in all this cases. Remember that we can apply it directly by splitting the terms of the known side in columns. Let's look at the next example.

Example 6.13 *Superposition*

Let us now write the previous example looking for the individual source contributions to the node voltages. Each term in the right hand vector of (6.18) is a sum of the form $b(2) + 5a$, where $b = 0$ in the first and third equation. The first term shows the contribution of the 2 mA source to the term, the second one that of the 5 V source. Writing for superposition means the individual terms in columns, writing the equation now in the form

$$
\begin{array}{c}
\text{Node1} \\
\text{Node2} \\
VCVS
\end{array}
\begin{bmatrix}
\frac{1}{8}+\frac{1}{5}+\frac{1}{2}+\frac{1}{6} & 0 & -\frac{1}{2}-\frac{1}{6} \\
0 & \frac{1}{1.5}+\frac{1}{12.5}+\frac{1}{4} & -\frac{1}{1.5} \\
4 & 0 & 1
\end{bmatrix}
\begin{bmatrix} V_1 \\ V_2 \\ V_3 \end{bmatrix}
=
\quad (6.19)
$$

$$
\begin{array}{c}
\text{Node1} \\
\text{Node2} \\
VCVS
\end{array}
\begin{bmatrix}
\text{``}I_n\text{''} & -5\,Y_{k4}\text{''} \\
0 & \frac{5}{8} \\
2 & \frac{5}{4} \\
0 & 0
\end{bmatrix}
$$

Once we have the correct settings, we can proceed as necessary, as it was done in example 6.7.

6.6.2 "Hybrid" reduced MNA

Section 6.5 deals with equations when we want all the unknown currents. On the other hand, the previous two subsection applies when we don't want any current, except when forced by the circuit itself. Well, in particular for our purposes, we could apply the principles of reduction only in a partial form so that we keep in the system specific unknown currents of interest. In later chapters we will see why these situations are important. Even more, fundamental for an efficient use of the calculator.

Whenever we need one or more specific currents, what we to do is to also take the equations where those currents appear. The following example illustrates this remark.

Example 6.14 *Take again the circuit from Fig. 6.25. If we want, for any reason, the current of the independent voltage source, all we have to do is to include the equation of node 4 in the set. We could also work example 6.12 and solve equation of node 4 separately. But then, what do we want the calculator for?*

6.6. "REDUCING" NUMBER OF EQUATIONS 129

$$\begin{array}{c} \\ Node1 \\ Node2 \\ Node4 \\ VCVS \end{array} \begin{bmatrix} V_1 & V_2 & V_3 & I_x \\ \frac{1}{8}+\frac{1}{5}+\frac{1}{2}+\frac{1}{6} & 0 & -\frac{1}{2}-\frac{1}{6} & 0 \\ 0 & \frac{1}{1.5}+\frac{1}{12.5}+\frac{1}{4} & -\frac{1}{1.5} & 0 \\ -\frac{1}{8} & -\frac{1}{4} & 0 & -1 \\ 4 & 0 & 1 & 0 \end{bmatrix} \begin{bmatrix} V_1 \\ V_2 \\ V_3 \\ I_x \end{bmatrix} =$$

$$\begin{array}{c} Node1 \\ Node2 \\ Node4 \\ VCVS \end{array} \begin{bmatrix} "I_n - 5\,Y_{k4}" \\ \frac{5}{8} \\ 2+\frac{5}{4} \\ -\frac{1}{8}-\frac{1}{4} \\ 0 \end{bmatrix}$$

(6.20)

We could work the problem also with superposition in mind.

As the reader might imagine, it is possible to program simplified nodal analysis. I leave this exercise to the interested reader.

6.6.3 Working with Operational Amplifiers

I close this chapter with comments and examples of nodal analysis for circuits containing ideal operational amplifiers. I could dedicate one chapter to this type of circuits or extend the chapter for several more pages. At this point, however, let us illustrate the main ideas with one circuit. The process I present is oriented toward hand analysis writing the equations directly .... of course!

The reason why I deal with OA separately is that, in a certain way, we work again with a sort of "supernode", but this time applied to the potential nodes, not to the equations! Let's see what I mean next. The principles governing this procedure are the following:

OA output: Unless you are interested on the current at the output of an OA, you do not take an equation at the output node.

OA input nodes potentials: All nodes connected through a chain of OA inputs are assigned the same variable. If one of the terminals is connected to a known voltage – including ground –, it is worked as a known voltage.

Equations at nodes: They follow the same principles as before, except that you should not forget to use the same variable, hence the same column, for all nodes connected by OA inputs.

Let us work an example using symbolic values so we can concentrate on the above rules. After that we work a numerical example.

Example 6.15 *The circuit in Fig. 6.26 has two OA's. Write the nodal equations.*

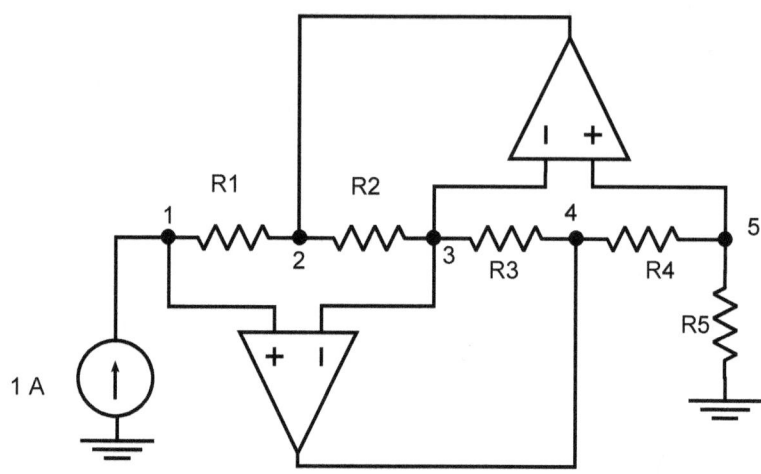

Figure 6.26: A circuit with Operational amplifiers

Nodes 1, 3 and 5 are all connected by OA inputs. Hence we use one variable, or one column, for the three potentials, because $V_1 = V_3 = V_5$. Let us use V_1. For the coefficients, apply the rule of (6.10) on page 92. For easy writing, I use G_j for $1/R_j$

$$\begin{array}{c} \\ \text{Node 1} \\ \text{Node 3} \\ \text{Node 5} \end{array} \begin{array}{c} V_1(V_3, V_5) \quad V2 \quad V4 \\ \begin{bmatrix} G_1 & -G_1 & 0 \\ G_2 + G_3 & -G_2 & -G_3 \\ G_4 + G_5 & 0 & -G_4 \end{bmatrix} \end{array} \begin{bmatrix} V_1 \\ V_2 \\ V_4 \end{bmatrix} = \begin{bmatrix} 1 \\ 0 \\ 0 \end{bmatrix}$$

Now a numeric example.

Example 6.16 *For the two OA circuit shown in Fig. 6.27, find the output voltage at node 2. The circuit has two inputs with known potentials Va and Vb.*

From the properties of ideal operational amplifiers, $V3 = Va$ and $V4 = Vb$. Since we are not interested on the output currents of the amplifiers, we take equations only at nodes 3 and 4. Applying the rules that we have seen in the chapter, and scaling resistances to kΩ units, we can write the equations as

$$\begin{array}{c} \\ \text{Node 3} \\ \text{Node 4} \end{array} \begin{array}{c} V1 \quad\quad V2 \\ \begin{bmatrix} -\frac{1}{1.2} & 0 \\ -\frac{1}{2.3} & -\frac{1}{4} \end{bmatrix} \end{array} \begin{bmatrix} V_1 \\ V_2 \end{bmatrix} = \begin{bmatrix} Va\,(=V3) & Vb\,(=V4) \\ -\frac{1}{0.8} - \frac{1}{1.2} - \frac{1}{0.5} & \frac{1}{0.8} \\ \frac{1}{0.8} & -\frac{1}{0.8} - \frac{1}{1.2} - \frac{1}{2.4} \end{bmatrix}$$

6.6. "REDUCING" NUMBER OF EQUATIONS

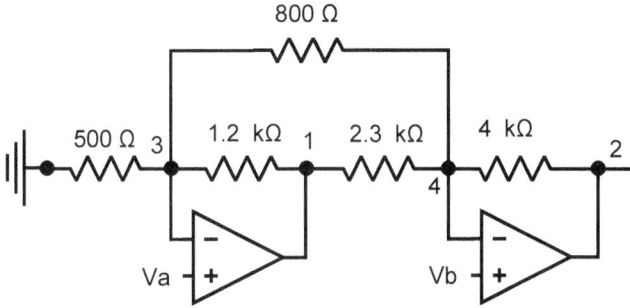

Figure 6.27: A circuit with Operational amplifiers

which we solve as

$$\begin{bmatrix} V_1 \\ V_2 \end{bmatrix} = \begin{bmatrix} \text{Va} & \text{Vb} \\ 4.9 & -1.5 \\ -13.522 & 12.609 \end{bmatrix}$$

Therefore, V2 = 12.609 Vb -13.522 Va.

CHAPTER 7

Loop Analysis

Loop analysis is as popular as nodal analysis among students. Perhaps more popular because it does not involve conductances. But then again, with calculators this should not be a problem.

Recall that a loop is a closed path in a circuit, where no node is touched twice. In planar networks drawn without crossing lines, a mesh is any loop which separates the plane in two half planes, and one of these does not contain any element. Consult your favorite textbook on this subject. Most examples in textbooks and also here, use planar examples. But that does not mean we are constrained to use meshes, or limited to planar circuits. A pair of examples will show this.

Following a textbook approach, I present the loop analysis method in two steps. First, following the traditional textbook approach working with circuits containing only resistances, known voltages and voltage dependent current sources. In the second stage, I include unknown voltages not dependent on currents. The starred sections are of theoretical nature and may be skipped by those with a good background.

7.1 Loop currents and loops selection

This section introduces the concepts around loop currents. The reader may skip it and come back if necessary.

A *loop current* is a mathematical entity envisioned as a current constrained to circulate within a loop, as illustrated in Fig. 7.1(a). If the loop is a mesh, the current is then also called *mesh current*. To set up the loop equations, we must select a set of B-N+1 independent loops where B is the number of elements and N the number of nodes in the circuit, including those for elements in series. In planar networks and working with meshes, it suffices to discard one of the meshes to have an independent set. For both planar and non planar circuits, there are several algorithms that allow us to select an independent set.

Any element must be in at least one loop, but it may belong to two or more

7.1. LOOP CURRENTS AND LOOPS SELECTION

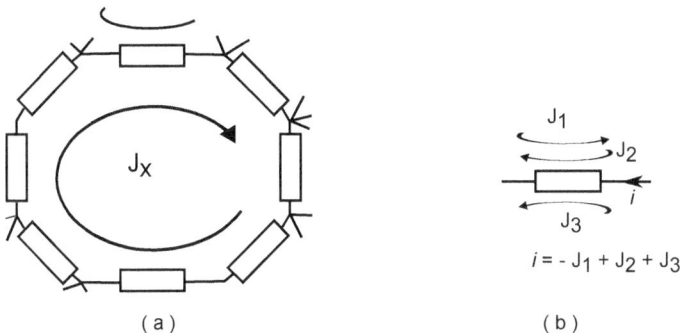

Figure 7.1: (a) Loop Current Concept. (b) Current in element as function of loop currents

loops. The current in the element is associated to the respective loop currents in the form illustrated by Fig. 7.1(b). Namely, the current in the element is the algebraic sum of the loop currents passing through it, with positive sign (+) if the loop current goes in the same direction as the element current, and minus sign (-) if it goes in the opposite direction,

$$i = \sum_{\substack{\text{same} \\ \text{direction}}} J_x - \sum_{\substack{\text{Opposite} \\ \text{direction}}} J_y \qquad (7.1)$$

In nodal analysis, once the reference node, or ground is selected, all equations are defined and potentials physical meaning. In loop analysis, however, we must select the set of loops and the direction of the respective currents, i. e., clockwise or counterclockwise. Fig. 7.2 illustrates this statement with a planar circuit with three meshes. Any two of these form an independent set. Hence, we have three possible independent sets. Yet, when direction is considered, there are twelve different cases of loop current selection. The figure shows only three of them.

Physical meaning for a loop current is possible only if there is an element in the loop that is not shared with other loops.

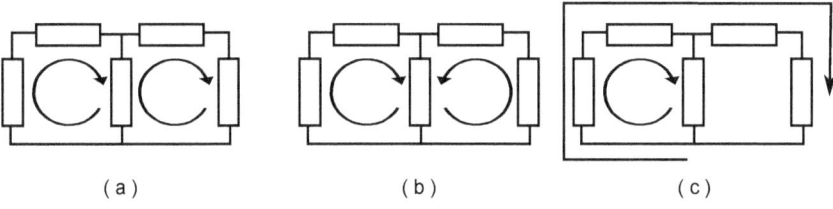

Figure 7.2: Two Mesh circuit: (a) The two internal meshes selected; (b) same meshes with different directions; (b) One internal mesh and the external mesh selected.

Fig. 7.3 illustrates another popular planar topology with four meshes. Not considering loop directions, there are 16 candidates sets of three independent loops. Including directions, the number becomes 128. The figure shows just two of these possibilities. Case (a) is the popular mesh selection. Notice that (b) includes a loop which is not a mesh and, moreover, there are elements which are shared by all three independent loops.

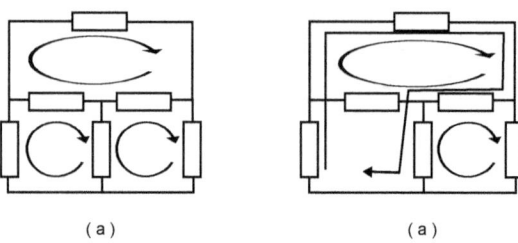

(a)   (a)

Figure 7.3: Another example of independent loops selection. Circuit with three loops

The basis for direct writing of loop equations is Kirchhoff's Voltage Law, which may be restated for our purposes as

Kirchhoff's Voltage Law: the sum of voltage drops in a loop is equal to the sum of voltage rises in the same loop.

Using this principle, the equation for a loop is written following the next statement:

Loop equation of loop x: Taking the direction of loop current J_x as reference, the sum of unknown voltage drops is equal to the sum of known voltage rises. A known voltage drop is taken as a negative known voltage rise. Similarly, an unknown voltage rise is take as a negative voltage drop. Remember that voltage drop means the current goes from the + terminal to the - terminal.

Fig. 7.4(a) shows an isolated loop x to illustrate the principle. V_a and V_b are assumed as unknown voltages. Also, in the direction of J_x, the voltage drop at resistance R_1, which is shared with loop currents J_1 and J_2, the voltage drop is $R_1(J_x - J_1 + J2)$, as illustrated in inset (b) of the figure. We work similarly with other resistances. Including the black boxes with unknown voltages, and the independent sources, the loop equation for this isolated loop becomes

$$R_1(J_x - J_1 + J_2) + R_2 J_x + R_3(J_x - J_3) + \ldots - V_b + V_a = E_1 - E_2 \qquad (7.2)$$

The unknown voltage drops may or may not be related to loop currents. The voltage in current sources are not related, for example, while current dependent voltage sources are related to loop currents.

7.2. CIRCUITS WITH RESISTANCES AND VOLTAGE SOURCES

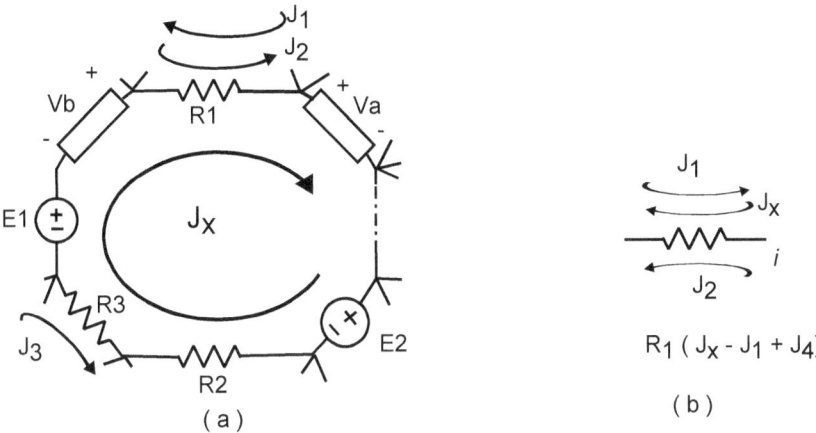

Figure 7.4: (a)Isolated loop x in circuit. (b) Voltage drop at R1 with respect to J_x

7.2 Circuits with resistances and voltage sources

First let us deal with circuits containing only resistances and voltage sources. Current sources may be transformed either by direct transformation or by first applying shifting of current sources. Our objective for these circuits is to write each loop k equation directly in the form

$$Z_{k1} J_1 + Z_{k2} J_2 + \cdots + Z_{km} J_m + \cdots = VL_k \quad (7.3)$$

Here

- $J_1, J_2, \ldots$ are the loop currents for loops 1, 2, ...
- Coefficients Z_{kj} are called *loop impedances* or *loop resistances*
- V_{Lk} is the loop voltage of loop k, i.e., the sum of voltage rises in the loop k

Each loop impedance Z_{mj} will consist of two parts: one due to resistances and another one arising from current-dependent voltage sources:

Z_{mj} = component from R's + component from current dependent voltages

Each component is independent of the other. We deal with these cases separately.
The set of all loop equations can be written in matrix form as

$$\mathbf{Zm\,Jm = V_L} \quad (7.4)$$

7.2.1 Circuits with only resistances and independent voltage sources

From (7.3), we see that in the vector of loop voltages $\mathbf{V_L}$, we have

$$VL_k = \sum (\text{known voltage rises in loop } k) \tag{7.5}$$

Fig. 7.5(b) illustrates how a source E contributes to the vector **VL**. Let us now consider the resistances.

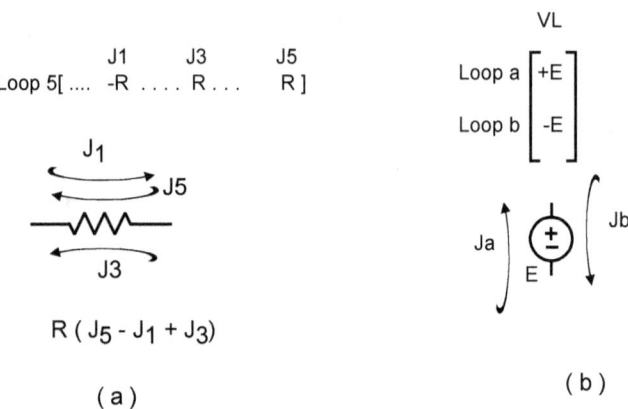

(a) (b)

Figure 7.5: (a) Contribution of Resistance to **Zm**. (b) Contribution of voltage source to **VL**

Looking at Fig. 7.4(b), we can state the following fact for resistances:

For each resistance included in loop k, and shared with loops q, p, ... there will be a voltage drop in the loop of the the form

$$(J_k \pm J_q \pm J_p \pm \ldots)R \tag{7.6a}$$

Therefore,

$$\text{the coefficient of } J_k \text{ is } = \sum \text{resistances in loop } k. \tag{7.6b}$$

and

$$\text{The coefficient of } J_p \text{ is } = \pm \sum \text{resistances shared by loops } k \text{ and } p \tag{7.6c}$$

where the sign + is taken if both loop currents flow through the resistance in the same direction, and the sign - in the opposite direction.

Summarizing, for the resistances contribution to coefficients of loop currents:

$$Z_{kj} = \begin{cases} +\sum \text{Resistances in loop } m & \text{for } j = k \\ \pm \sum \text{Resistances shared by loops } m \text{ and } j & \text{for } j \neq m \end{cases} \tag{7.7}$$

Fig. 7.5(a) illustrates how a resistance R contributes to the equation of a loop, which was labeled 5 for the purpose of illustration.

7.2. CIRCUITS WITH RESISTANCES AND VOLTAGE SOURCES

Let us bring two examples applying (7.7). In both examples, equations are written in matrix form. The first one goes step by step, showing clearly the application of the criteria. To stress the process, a non numerical example is used. The second one is done straightforward by applying the rules, leaving the step by step verification to the reader.

Example 7.1 *Let us write the loop equations for the circuit of Fig. 7.6, with the three loops selected as shown.*

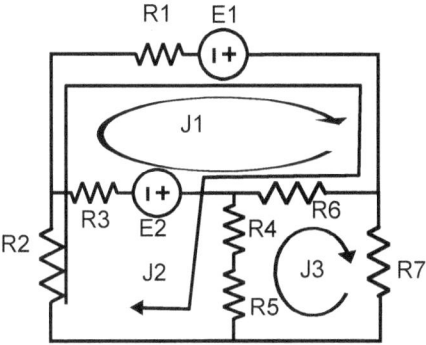

Figure 7.6: Circuit for Step by Step example

To proceed to the individual equations, let us consider the three loops separately, as illustrated in Fig. 7.7.
(1) Equation for Loop 1, shown in detail by inset (a):

(1.1) The resistances contained in loop 1 are R_1, R_3, and R_6. The coefficient for J_1 is therefore $R_1 + R_3 + R_6$.

(1.2) The resistances shared by loop 1 and loop 2 are R_1 and R_6. The respective loop currents go in the same direction through these resistances. Hence, the coefficient for J_2 in this equation is $R_1 + R_6$.

(1.3) The resistance shared by loop 1 and loop 3 is R_6. The respective loop currents go in the opposite direction through this resistance. Hence, the coefficient for J_3 in this equation is $-R_6$.

(1.VL) Loop 1 contains two independent voltage sources. The direction of J_1 is such that it traverses E_1 in a voltage rise direction, i. e., from the minus (−) to the plus sign. For E_2, the direction is in drop sense. Therefore, the known right hand side should be $E_1 - E_2$.

The equation for loop 1 is therefore

$$(R_1 + R_3 + R_6) J_1 + (R_1 + R_6) J_2 - R_6 J_3 = E_1 - E_2 \qquad (7.8a)$$

(2) Equation for Loop 2, shown in detail by inset (b):

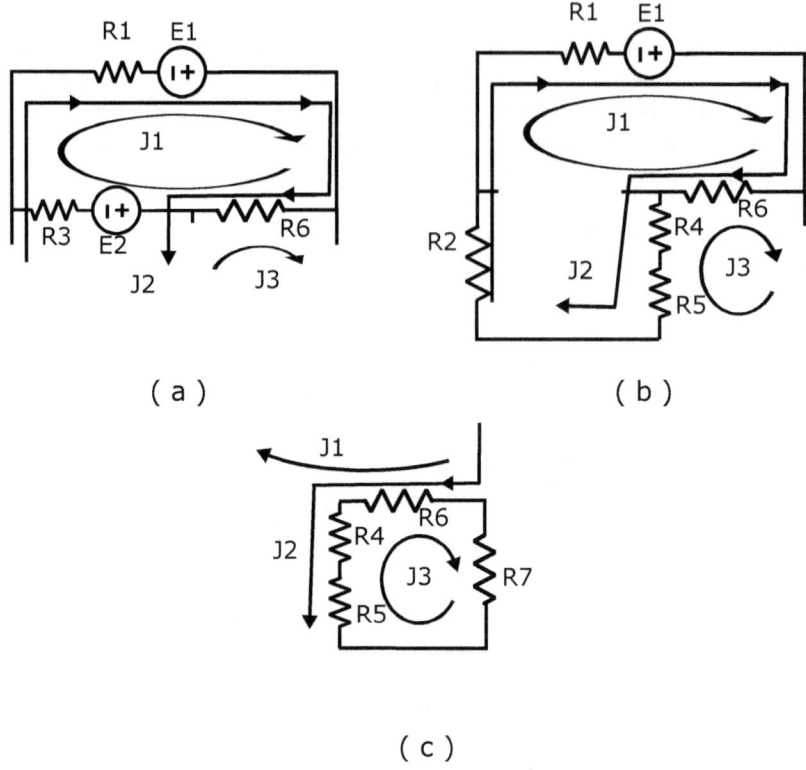

Figure 7.7: (a) Loop 1; (b) Loop 2; (c) Loop 3

(2.1) The resistances shared by loop 1 and loop 2 are R_1 and R_6. The respective loop currents go in the same direction through these resistances. Hence, the coefficient for J_1 in this equation is $R_1 + R_6$.

(2.2) The resistances contained in loop 2 are $R_1, R_2, R_4, R_5,$ and R_6. The coefficient for J_2 is therefore $R_1 + R_2 + R_4 + R_5 + R_6$.

(2.3) The resistances shared by loop 2 and loop 3 are $R_4, R_5,$ and R_6. The respective loop currents go in the opposite direction through these resistances. Hence, the coefficient for J_3 in this equation is $-(R_4 + R_5 + R_6)$.

(2.VL) Loop 2 contains one independent voltage source E_1. The direction of J_2 through E_1 is in a voltage rise sense. Thus, the known right hand side should be E_1.

The equation for loop 2 is therefore

$$(R_1 + R_6) J_1 + (R_1 + R_2 + R_4 + R_5 + R_6) J_2 - (R_4 + R_5 + R_6) J_3 = E_1 \quad (7.8b)$$

(3) Equation for Loop 3, shown in detail in inset (c):

7.2. CIRCUITS WITH RESISTANCES AND VOLTAGE SOURCES

(3.1) The resistance shared by loop 3 and loop 1 is R_6. The respective loop currents go in the opposite direction through this resistance. Hence, the coefficient for J_1 in this equation is $-R_6$..

(3.2) The resistances shared by loop 3 and loop 2 are $R_4, R_5,$ and R_6. The respective loop currents go in the opposite direction through these resistances. Hence, the coefficient for J_2 in this equation is $-(R_4 + R_5 + R_6)$.

(3.3) The resistances contained in loop 3 are $R_7, R_4, R_5,$ and R_6. The coefficient for J_3 is therefore $R_1 + R_4 + R_5 + R_6 + R_7$.

(1.VL) Loop 3 contains no independent voltage source. Therefore, the known right hand side is 0.

The equation for loop 2 is therefore

$$-R_6 J_1 - (R_4 + R_5 + R_6) J_2 + (R_4 + R_5 + R_6 + R_7) J_3 = 0 \qquad (7.8c)$$

In matrix form, we have

	J_1	J_2	J_3
Loop1	$R_1 + R_3 + R_6$	$R_1 + R_6$	$-R_6$
Loop2	$R_1 + R_6$	$R_1 + R_2 + R_4 + R_5 + R_6$	$-(R_4 + R_5 + R_6)$
Loop3	$-R_6$	$-(R_4 + R_5 + R_6)$	$R_4 + R_5 + R_6 + R_7$

$$\times \begin{bmatrix} J_1 \\ J_2 \\ J_3 \end{bmatrix} = \begin{bmatrix} VL \\ E_1 - E_2 \\ E_1 \\ 0 \end{bmatrix}$$

(7.9)

As one would expect, the coefficient matrix in the example is symmetrical.

Remark: A common mistake among students is the belief that the circuit element voltages and currents will be different depending on how we select the loops and the direction of the loop current, clockwise or counterclockwise. The misunderstanding arises from the fact that loop currents are considered as "physical" circuit variables. Although this is a common notation and also sometimes they have physical meaning, rigourously speaking, they are not circuit variables but mathematical entities to find the magnitudes of interest, which are the voltages and currents in the elements. These do not depend on the selection of loops, or selection of ground in the nodal analysis. Let us illustrate with the same circuit as before, but now assigning numerical values to elements.

Example 7.2 *Find all the voltage and currents in the circuit of Fig. 7.8 using two different sets of loop currents.*

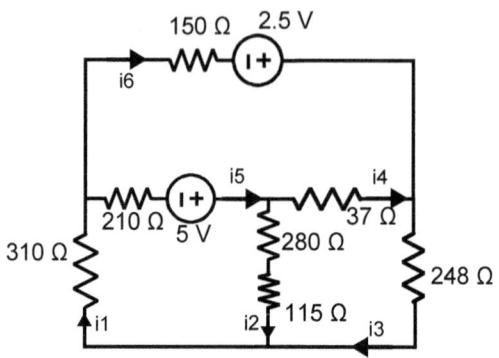

Figure 7.8: Now with numerical values

Fig. 7.9(A) show as selection of independent loop currents the same set used in the previous example. For the other set, we use mesh currents. This is a planar circuit that has four meshes so any three mesh currents constitute an independent set. We pick the three internal meshes, as shown in Fig. 7.9(B).

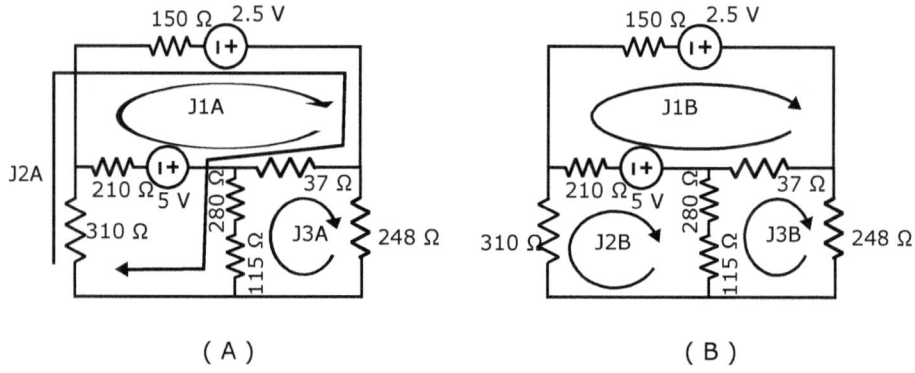

Figure 7.9: Two different selections, (A) y (B), of loops.

Before proceeding to establish and solve the loop equations, let us write down how the individual branch currents will be determined in each case. This is done in the following table:

	i_1	i_2	i_3	i_4	i_5	i_6
Case A:	J2A	J2A - J3A	J3A	J3A-J1A-J2A	-J1A	J1A + J2A
Case B:	J2B	J2B - J3B	J3A	J3B - J1B	J2B - J1B	J1B

From the results for i_1 and i_3, we should expect J2A = J2B and J3A = J3B.

7.2. CIRCUITS WITH RESISTANCES AND VOLTAGE SOURCES

Now let us solve for the selection A. The equations are, following the procedure established before,

$$
\begin{array}{c} \\ Lp1A \\ Lp2A \\ Lp3A \end{array}
\begin{bmatrix}
J_{1A} & J_{2A} & J_{3A} \\
150 + 37 + 210 & 150 + 37 & -37 \\
150 + 37 & 310 + 150 + 37 + 280 + 115 & -(37 + 280 + 115) \\
-37 & -(37 + 280 + 115) & 280 + 115 + 37 + 248
\end{bmatrix} \times
$$

$$
\begin{bmatrix} J_{1A} \\ J_{2A} \\ J_{3A} \end{bmatrix} = \begin{bmatrix} V_L \\ 2.5 - 5. \\ 2.5 \\ 0 \end{bmatrix}
$$

(7.10)

For the selection B, we have

$$
\begin{array}{c} \\ Lp1B \\ Lp2B \\ Lp3B \end{array}
\begin{bmatrix}
J_{1B} & J_{2B} & J_{3B} \\
150 + 37 + 210 & -210 & -37 \\
-210 & 310 + 210 + 280 + 115 & -(280 + 115) \\
-37 & -(280 + 115) & 280 + 115 + 37 + 248
\end{bmatrix} \times
$$

$$
\begin{bmatrix} J_{1B} \\ J_{2B} \\ J_{3B} \end{bmatrix} = \begin{bmatrix} V_L \\ 2.5 - 5.0 \\ 5.0 \\ 0 \end{bmatrix}
$$

(7.11)

Let us store matrices and vectors as za and va for the first set, and zb and vb for the second. Store solutions as xa and xb, respectively Then

$$\text{za} \wedge \boxed{(-)} 1 * \text{va} \boxed{\text{STO}\blacktriangleright} \text{ xa} \rightarrow \begin{bmatrix} -8.990 \text{ E-3} \\ 6.428 \text{ E-3} \\ 3.595 \text{ E-3} \end{bmatrix}$$

Similarly,

$$\text{zb} \wedge \boxed{(-)} 1 * \text{vb} \boxed{\text{STO}\blacktriangleright} \text{ xb} \rightarrow \begin{bmatrix} -2.562 \text{ E-3} \\ 6.428 \text{ E -3} \\ 3.595 \text{ E-3} \end{bmatrix}$$

As expected, J2A = J2B = 6.428 mA *and* J3A = J3B = 3.59 mA. *Hence, currents* i_1, i_2, *and* i_3 *have the same value in both systems. Checking for the other cases:*

For i_4: xa[3,1] - xa[1,1] - xa[2,1] $\boxed{\text{ENTER}}$ → 6.157 E-3
and xb[3,1] - xb[1,1] $\boxed{\text{ENTER}}$ → 6.157 E-3, i. e., i_4 = 6.157 mA.

For i_5: xb[2,1] - xb[1,1] $\boxed{\text{ENTER}}$ → 8.990 E-3, i. e., i_5 = 8.990 mA., *same as* -xa[1,1] (-J1A)

For i_6: xa[1,1] + xa[2,1] $\boxed{\text{ENTER}}$ → -2.562 E-3, i. e., i_6 = -2.562 mA., *same as* xb[1,1] (J1B)

Scaling units

We can introduce the resistances in kilohms units, and then all currents can be interpreted in miliampere units.

7.2.2 Current-controlled voltage sources

Let us now introduce the controlled voltage sources into the game. We assume they are current controlled. For the specific case in which they are voltage controlled but the controlling voltage is related to current, then we simply apply this relation.

The general situation is depicted in Fig. 7.10, in which the source is shown belonging to two loops, the currents being Jp and Jq. Of course, just one or more than two are possible, but that does not alter the general discussion. The controlling current is expressed here as the difference of two loop currents as Ja - Jb.

The equations that will be modified by the source are for those loops to which the the source belongs. The way in which this modification takes place is illustrated in the same figure. Namely, the trans resistance of control r will add to or subtract from the coefficients of the respective controlling loop currents.

We illustrate this remark using the situation shown in Fig. 7.10. For loop p, seeing that current Jp traverses the source in the voltage drop direction, and thus positive, we write the equation in the form

$$Z'_{p1} J_1 + \ldots + Z'_{pa} J_a + \ldots + Z'_{pb} J_b + \ldots + r(J_a - J_b) = VL_p \qquad (7.12a)$$

where the prime in the notation is to emphasize that the coefficients have been derived taking into account resistances only, with the rules of the previous section.

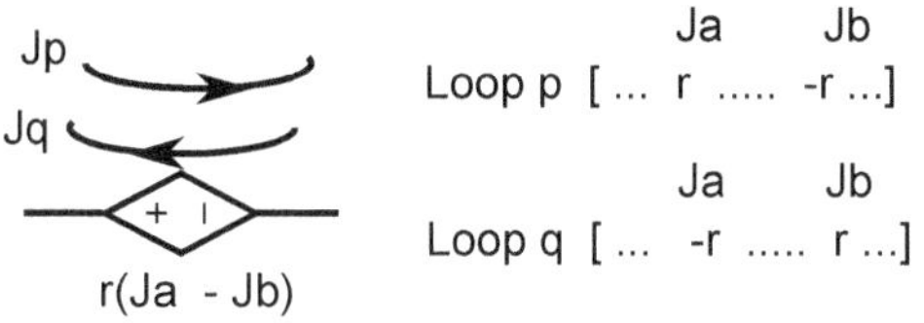

7.2. CIRCUITS WITH RESISTANCES AND VOLTAGE SOURCES

Figure 7.10: CCVS affects the loop equations in which it is embedded

Similarly, for loop q,

$$Z'_{q1} J_1 + \ldots + Z'_{qa} J_a + \ldots + Z'_{qb} J_b + \ldots - r(J_a - J_b) = VL_q \qquad (7.12b)$$

Rearranging the equations with Algebra,

$$Z'_{p1} J_1 + \ldots + (Z'_{pa} + r) J_a + \ldots + (Z'_{pb} - r) J_b + \ldots = VL_p \qquad (7.13a)$$

$$Z'_{q1} J_1 + \ldots + (Z'_{qa} - r) J_a + \ldots + (Z'_{qb} + r) J_b + \ldots = VL_q \qquad (7.13b)$$

When you set up the equations, you may either do it in two steps, starting with the set (7.12), or with practice directly to the final forms (7.13).

Let us work an example. I take advantage of this example to illustrate partially how the program for mesh analysis discussed in section 7.3 is planned.

Example 7.3 *Let us find the voltage Vo in the circuit of Fig. 7.11 using mesh analysis.*

We proceed in two steps. First, imagine that the dependent source is not included, considering it to be a short circuit, and set up the matrices. With the dependent source subtituted with a short circuit, **Zm** *is*

	J_1	J_2	J_3	J_4
Loop1	10000 + 600	−10000	−600	0
Loop2	−10000	10000 + 500 + 1200	−1200	0
Loop3	−600	−1200	600 + 1200 + 150	−150
Loop4	0	0	−150	150 + 1500

and

$$\mathbf{VL} = \begin{bmatrix} 3 \\ 0 \\ 0 \\ 0 \end{bmatrix}$$

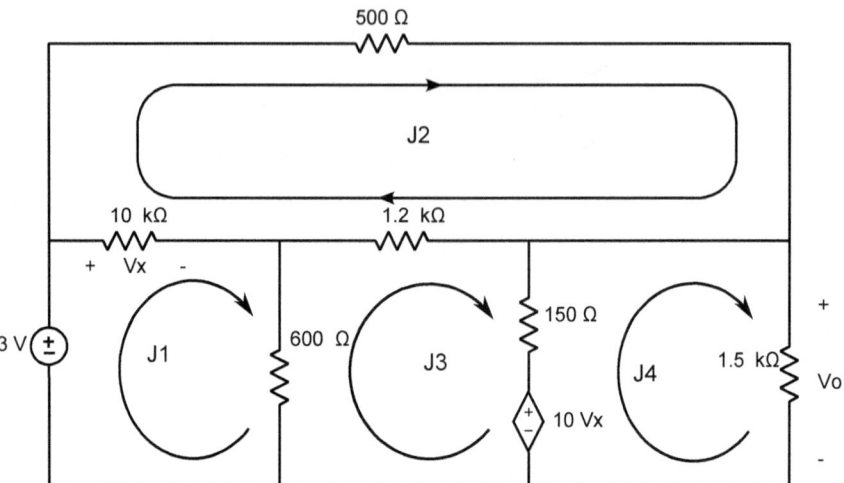

Figure 7.11: Example for Mesh Analysis with Current Controlle Voltage Source.

Let us store the matrices as `zm` and `vl` in our calculator. We are now ready to consider the dependent voltage source. This one is actually a voltage controlled source. But since Vx = 10000 (J1-J2), we can express it in terms of loop currents as 10(10E3)(J1-J2)=10E4(J1 - J2).

Looking at the circuit, we see that this dependent source is in drop sense for J3 and rise direction for J4. Using as reference what has been said before, in equations of J3 and J4, the coefficients for J1 and J2 will be affected as shown in (7.13).

Therefore, the final matrix is as shown below, where I put the source contribution in bold, and using E notation:

$$\mathbf{Zm} = \begin{array}{c} \\ Lp1 \\ Lp2 \\ Lp3 \\ Lp4 \end{array} \begin{bmatrix} J_1 & J_2 & J_3 & J_4 \\ 10000 + 600 & -10000 & -600 & 0 \\ -10000 & 10000 + 500 + 1200 & -1200 & 0 \\ -600 + \mathbf{10E4} & -1200 - \mathbf{10E4} & 600 + 1200 + 150 & -150 \\ -\mathbf{10E4} & \mathbf{10E4} & -150 & 150 + 1500 \end{bmatrix}$$

In the calculator, having already the matrix without dependent source, we proceed to get the final matrix `zm` with the following four entries on the command line:

`zm[3,1] + 10E4` STO▶ `zm[3,1]` → 99400.

7.2. CIRCUITS WITH RESISTANCES AND VOLTAGE SOURCES

zm[3,2] - 10E4 [STO▶] zm[3,2] → -1.012E5

zm[4,1] - 10E4 [STO▶] zm[4,1] → -1.E5

zm[4,2] + 10E4 [STO▶] zm[4,2] → 1.E5

We are now ready to find the solution as

$$zm \wedge [(-)]1 * vl \; [\text{ENTER}] \to \begin{bmatrix} -6.61\text{E-}3 \\ -6.69\text{E-}3 \\ -10.19\text{E-}3 \\ 4.23\text{E-}3 \end{bmatrix}$$

We have therefore J4 = 4.23 mA. We multiply now this current by the resistance 1.5 kΩ to find the desired output voltage:

ans[4,1]*1.5E3 [ENTER] → 6.3455E0

Hence, Vo = 6.35 V.

A trick: Insert equation or change variable You are working with technology, a calculator. Why not include more information to do things faster? In the previous example, since J4 = Vo/1500, we could have change variable directly in the equations. Thus, our matrix, *including the CCVS*, would be

Zm =

	J_1	J_2	J_3	Vo
L1	$10000 + 600$	-10000	-600	0
L2	-10000	$10000 + 500 + 1200$	-1200	0
L3	$-600 + 100E3$	$-1200 - 100E3$	$600 + 1200 + 150$	$-150/1500$
L4	$0 - 100E3$	$0 + 100E3$	-150	$(150 + 1500)/1500$

and the result would be direct. Try it!

7.2.3 Indefinite mesh matrix

Let us limit ourselves to planar networks, and select all meshes as loops, including the external mesh. Furthermore, let all internal meshes be clockwise oriented, and the external mesh counterclockwise oriented. The resultant matrix **Zm** is called *indefinite mesh impedance matrix* (IMM). By extension, **Vm**, may be called indefinite mesh voltage vector. The IMM has many uses[1]. Here, the purpose is to use it to program mesh analysis.

---

[1] For further information the reader may consult the paper by Kiss, W.F. ; Gilson, R.A. , "On the formulation of the indefinite matrix", IEEE J. of Solid-State Circuits, vol. 3, No. 3, pp. 307-308, 1968

Both matrices **Zm** and **Vm** have the characteristic that the sum of the rows are 0. Moreover, in **Zm**, the sum of columns is also 0. As a consequence, its determinant is 0 an the set of equations does not have a solution. For our purpose, however, and important characteristic is that we can delete the column and row that correspond to a mesh in **Zm** and **Vm**. The system that results corresponds to the independent set of meshes that remain.

7.3 Programming mesh equations I

Let us program to generate the matrices **Zm** and **Vm** and solve for the mesh current vector **Jm** . The basic algorithm for circuits with characteristics mentioned so far, that is, with resistances, votage sources and current controlled voltage sources, is described in the following steps using indefinite mesh matrix. The mesh we want to drop is denoted as "N+1" for easy row and column deletion.

7.3.1 Pseudocode for program

The pseudo code given below assumes that you have at least one loop in your system.

1. **Number of meshes** Specify number of meshes N, excluding the one you drop. Denote the dropped mesh as N+1.
2. **Initialization** Initialize matrix **Zm** = 0 of order (N+1)x(N+1) as well as vector **Vm** of order (N+1)x1
3. **Resistance Subroutine** For each resistance of value R ($\neq 0, \infty$) belonging to meshes j and k do:

    1. $Zm[j,j] = Zm[j,j] + R$
    2. $Zm[k,k] = Zm[j,j] + R$
    3. $Zm[j,k] = Zm[j,k] - R$
    4. $Zm[k,j] = Zm[k,j] - R$

4. **Current controlled voltage source subroutine** If there are CCVS's, then, for each source with transresistance rm in drop direction for mesh j and rise direction for mesh k, and controlled by $(J_p - J_q)$ do:

    1. $Zm[j,p] = Zm[j,p] + rm$
    2. $Zm[j,q] = Zm[j,q] - rm$
    3. $Zm[k,p] = Zm[k,p] - rm$
    4. $Zm[k,q] = Zm[k,q] + rm$

5. **Creating definite Zm:** Delete both row and column (N+1) of **Zm**.
6. **Voltage source Subroutine** For each cource source of value V ($\neq 0, \infty$) in drop position for mesh j and rise position for mesh k do:

    1. $Vm[j,1] = Vm[j,1] - V$

7.3. PROGRAMMING MESH EQUATIONS I

2. $Vm[k, 1] = Vm[k, 1] + V$

7. Creating definite Vm: Delete both row (N+1) of **Vm**
8. Solve for Jm: If $|Zm| \neq 0$ then $\mathbf{Jm} = \mathbf{Zm}^{-1}\mathbf{Vm}$

If there are no voltage sources, you can use the program to get the matrix **Zm**. Let us now use the above steps to develop programs.

7.3.2 Preparing input for the program

The program receives the following inputs. Remember, you must include the outer mesh current in counter clockwise direction, and all internal mesh currents in clockwise direction.

1. **n**: A non zero number of independent meshes. The description includes all meshes, and the one to be dropped out is called mesh "**n+1**" in the following matrices

2. An $m \times 3$ matrix **R** for the m resistances in the circuit. Each row k consists of three items:
   - Element $(k, 1)$ is a non-zero and non-infinite resistance value,
   - Elements $(k, 2)$ and $(k, 3)$ are the two meshes to which it belongs.

3. Variable **vs**. This variable is 0 if there are no independent voltage sources, which is the case if you are only interested on **Zm**. Otherwise, is a $p \times 3$ matrix **vs** for the p voltage sources in the circuit. Each row k consists of three items:
   - Element $(k, 1)$ is the non-infinite voltage source value. It may be 0.
   - Element $(k, 2)$ is the mesh number where the current is in voltage drop sense through the source,
   - Element $(k, 3)$ is the mesh number where the current is in voltage rise sense through the source.

4. Input **rs**. This value is 0 if the circuit contains no current controlled voltage sources. Otherwise, it is a $p \times 5$ matrix. The source $rm(Jp - Jq)$ is defined in row k as follows:
   - Element $(k, 1)$ is a non-infinite transresistance rm value.
   - Element $(k, 2)$ is the mesh where current is in voltage drop sense (entering through the + terminal),
   - Element $(k, 3)$ is the mesh where current is in voltage rise sense (entering through the - terminal),
   - Element $(k, 4)$ is the mesh number for Jp.
   - Element $(k, 5)$ is the mesh number for Jq.

Fig. 7.12 illustrates the rows for each of the input matrices,

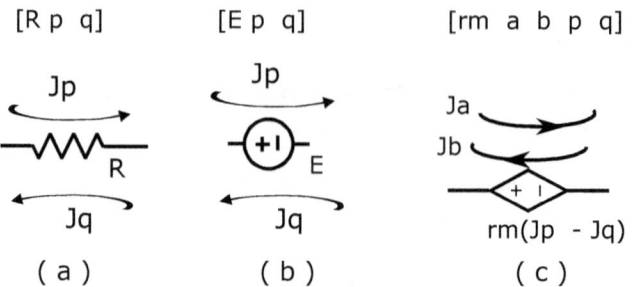

Figure 7.12: Illustrating rows for matrices (a) **r**, (b) **vs**, and (c) **rs**

7.3.3 Program for the calculators

Fig. 7.13 shows a program in these environments. The lines are commented on the right column which are not part of the program, although you can include them if you wish. Except for the case of checking if the circuit description is valid, I have not attempted any extra feature like data checking or other advisable programming precautions for a general case. I wrote the program for personal use and I assume that at least I will not provide wrong data.

The program returns three matrices: The loop impedance matrix **zm**, the loop voltage vector **vm**, and the vector of loop currentes **jm**. These results are in memory once you run the program.

Example 7.4 *Let us illustrate using the circuit worked in example 7.3. In this circuit the four inner meshes were selected. This means that the extra mesh is the outer mesh, which goes in the counterclock direction. This is illustrated in Fig. 7.14.*

The inputs for the program to solve for this circuit, with the selection shown, are:

$$\text{n} = 4; \quad \text{r} = \begin{bmatrix} 10E3 & 1 & 2 \\ 600 & 1 & 3 \\ 1200 & 2 & 3 \\ 150 & 3 & 4 \\ 1500 & 4 & 5 \\ 500 & 3 & 5 \end{bmatrix}; \quad \text{vs} = [3,5,1]; \quad \text{rs} = [100E3, 3, 4, 1, 2]$$

7.4. LOOP ANALYSIS INCLUDING CURRENT SOURCES

1	:loop1(n,r,vs,rs)	
2	:Pgrm	
3	:Local a,b,c,d,e,t	
4	:newMat(n+1,n+1)→ zm	Initializ indefinite mesh matrix
5	:newMat(n+1,1)→ vm	and indefinite Vm vector
6	:For t,1,rowDim(r)	Start Zm construction with resistances
7	:r[t,1]→a::r[t,2]→b	For simpler notation in writing: a=R;
8	:r[t,3]→c	R belongs to meshes b and c
9	:zm[b,b]+a→zm[b,b]	Updating Zm_{bb}
10	:zm[c,c]+a→zm[c,c]	Update Zm_{cc}
11	:zm[b,c]-a→zm[b,c]	Updating Zm_{bc} and
12	:zm[c,b]-a→zm[c,b]	Updating Zm_{cb}
13	:EndFor	End Resistance subroutine
14	:	Blank line for easy reading
15	:If rs/=0 Then	Case when there are VCCS's
16	:For t,1,rowDim(rs)	Start introduction of dependent sources
17	:rs[t,1]→a::rs[t,2]→b	For simpler notation in writing: a=rm;
18	:rs[t,3]→c	drop in mesh b, raise in mesh c
19	:rs[t,4]→d: rs[t,5]→e	controlled by Jmd - Jme
20	:zm[b,d]+a→zm[b,d]	Updating Zm_{bd} and
21	:zm[b,e]-a→zm[b,e]	Updating Zm_{be} and
22	:zm[c,d]-a→zm[c,d]	Updating Zm_{cd} and
23	:zm[c,e]+a→zm[c,e]	Updating Zm_{ce} and
24	:EndFor	End reading rs
25	:EndIf	End case for dependent sources
26	:	Blank line for easy reading
27	:subMat(zm,1,1,n,n)→ zm	Create definite zm
28	:	Blank line for easy reading
39	:If vs/=0 Then	Case for independent sources
30	:For t,1,rowDim(vs)	Introduce independent sources
31	:vs[t,1]→a::vs[t,2]→b	For simpler notation in writing: a=I;
32	:vs[t,3]→c	drop in mesh b, raise in mesh c
33	:vm[b,1]-a→vm[b,1]	Updating Vm_b
34	:vm[c,1]+a→vm[c,1]	Updating Vm_c
35	:EndFor	End Independent currents subroutine
36	:	Blank line for easy reading
37	:	Blank line for easy reading
38	:subMat(vm,1,1,n,1)→ vm	Definite vm
39	:If det(zm)/=0 Then	If circuit is valid
40	:zm∧-1*im→ jm	Solving for mesh current vector jm
41	:Else	Otherwise
42	:Disp ''Circuit description non valid''	Message for invalid description
43	:EndIf	End mesh currents
44	:EndIf	End case of independent sources
45	:EndPrgm	Exit Program

Figure 7.13: Example of a program for mesh analysis (resistances and current sources). Remember that → in the program stands for STO▶

7.4 Loop Analysis including Current Sources

When there are current sources in your circuit you will have unknown voltages which do not depend on your loop currents. Therefore, the number of variables increases. Each current source introduces an unknown voltage and also an equation

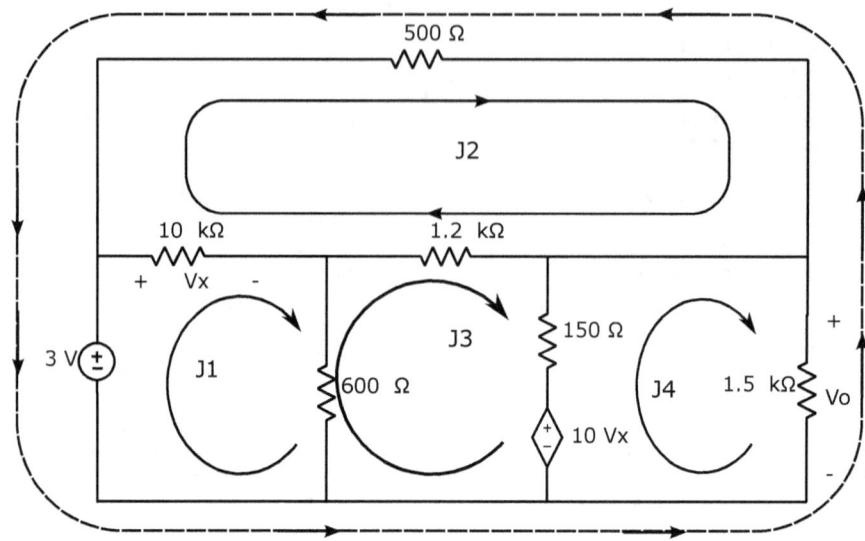

Figure 7.14: Circuit to illustrate inputs to program mesh1.

or a known value and the system is solvable[2].

You have two options. Either you take a reduced number of equations not involving the unknown voltages, or else you include those voltages in the system. The firs option is the preferred one when solving with pencil and paper. Since we are using calculators which help us solve the equations easily, the way we proceed is a matter of choice.

7.4.1 Modified Loop Analysis

When you want all the unknown voltages as well, you can include them in the equations. You don't need to select loops in a special way. Just do it the way you prefer! The result has been called by some authors modified loop analysis. An advantage is that it can be also programmed when it come to meshes. I leave that task to the reader and just give here the hints.

Assume there are N meshes (loops) and B current sources. The equations have the form $\mathbf{Z_{mla}J_{ma}} = \mathbf{V_{mla}}$, where matrices and vectors can be partitioned, so that the final form is

$$\begin{bmatrix} \mathbf{Z_m} & \mathbf{V_x} \\ \mathbf{IEQ} & \mathbf{0} \end{bmatrix} \begin{bmatrix} \mathbf{J}_m \\ \mathbf{V}x \end{bmatrix} = \begin{bmatrix} \mathbf{V}L \\ \mathbf{I}_s \end{bmatrix} \qquad (7.14)$$

where

$\mathbf{Z_m}$ of order N x N, $\mathbf{VL}$ of order N x 1 are the same matrices already introduced. $\mathbf{J_m}$ is the vector of unknown mesh currents

---

[2]Remember that we always assume well stated systems for our purposes.

7.4. LOOP ANALYSIS INCLUDING CURRENT SOURCES

$\mathbf{V_x}$ is an N x B matrix of for the unknown voltages in sources, and $\mathbf{IEQ}$ is the B x N matrix defining the current sources equations. The unknown voltage in column j of $\mathbf{V}_x$ stands for that of the current for row j of $\mathbf{IEQ}$. This colum j has +1 in the row for mesh current J_p if the current runs in voltage drop direction, -1 if it runs in voltage rise direction, and 0 otherwise.

The system $\mathbf{IEQ\,V_x = I_s}$ is what defines the set of current sources.

The following example by hand will illustrate the structure. Remember: *In the unknown side, an unknown voltage drop has coefficient +1 (positive) and an unknown voltage rise has coefficient -1 (negative)!*

Example 7.5 *To illustrate the process, let us take once more the same circuit of Fig. 7.16, which is reproduced in Fig. 7.15 with meshes as usually taken by students. In this circuit, let us find the power generated by each of the three sources.*

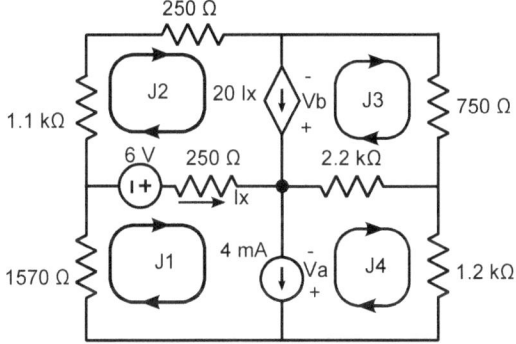

Figure 7.15: Modified mesh analysis for the circuit of Fig. 7.16

Before proceeding to set up the equations, we need to have clear two facts:

a) How are we going to solve the problem once we have the solution for the equations. Remember, the equations will provide us the values of the loop currents and the unknown voltages, not the power calculations that are asked for.

b) In setting up the equations, what are the relations introduced by the current sources.

Regarding the item a), the power calculations can be written in terms of the variables in the equations as

$$P_{V-source} = 6\times(J_1-J_2); \quad P_{I-source} = (4\times10^{-3})V_a; \quad P_{CCCS} = V_b\times(J_2-J_3) \quad (7.15)$$

As for item b), the equations defined by the current sources are:

$$J_1 - J_4 = 4\times10^{-3}; \quad \text{and} \quad J_2 - J_3 = 20(J_1 - J_2), \quad \text{or} \quad -20J_1 + 21J_2 - J_3 = 0$$

Now, set up the equations. The matrices involved are as follows:

$$\mathbf{Zm} = \begin{array}{c} \\ Lp1 \\ Lp2 \\ Lp3 \\ Lp4 \\ CCCS \\ Isrc \end{array} \begin{bmatrix} J_1 & J_2 & J_3 & J_4 & V_a & V_b \\ 250+1570 & -250 & 0 & 0 & -1 & 0 \\ -250 & 1100+250+250 & 0 & 0 & 0 & -1 \\ 0 & 0 & 2200+750 & -2200 & 0 & 1 \\ 0 & 0 & -2200 & 2200+1200 & 1 & 0 \\ -20 & 21 & -1 & 0 & 0 & 0 \\ 1 & 0 & 0 & -1 & 0 & 0 \end{bmatrix} \quad (7.16)$$

$$\mathbf{VL'} = \begin{bmatrix} 6 \\ -6 \\ 0 \\ 0 \\ 0 \\ 0.004 \end{bmatrix} \quad (7.17)$$

Can you identify the submatrices? Now solve the system. For easiness, we show the result in transposed form:
(zm∧ (-) 1* vl) STO▶ xm ENTER →
[372.77E-6 265.86E-6 -1.873E-3 372.77E-6 -5.388E-0 6.3454E-0]^T

The vector elements are are, respectively, J1, J2, J3, J4, Va, and Vb, respectively. We can now solve for the required power calculations of (7.15).

For $P_{V-source} = 6 \times (J_1 - J_2)$: 6 * (xm[1,1] - xm[1,2]) → 641.65E-6
For $P_{I-source} = (4 \times 10^{-3})V_a$: 4E-3 * xm[1,5] → -21.552E-3
For $P_{CCCS} = V_b \times (J_2 - J_3)$: xm[1,6] * (xm[1,2] - xm[1,3]) → 13.572E-3

Therefore, the voltage source generates 641.65 μW, the dependent source generates 13.572 mW, but the independent current source absorbs 21.552 mW.

7.4.2 Reducing the number of equations

This method is based on the principle, already stated before, that if an unknown variable appears in only one equation, than the equation may be taken separately. If the variable is not of interest, the equation is not solved.

In the case of loop equations, the isolation of the unknown voltage is achieved by selecting loops in such a way that one and only one loop goes through the

7.4. LOOP ANALYSIS INCLUDING CURRENT SOURCES

current source. If the source is independent, we can take the loop current as a known value[3]. If it is dependent, then we need to add the equation of the source to the system.

Let us look at an example.

Example 7.6 *Let us set up the equations for the circuit of Fig. 7.16. This circuit contains two current sources. Our objective is to write a set of equations which does not include the voltages at those sources, V_a and V_b.*

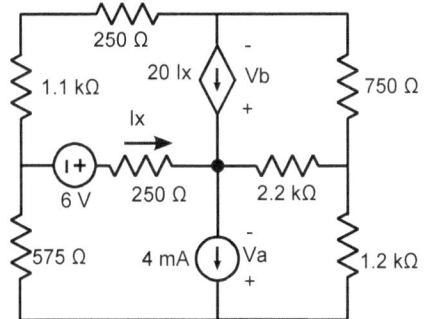

Figure 7.16: Circuit with current sources

We start by drawing loops in a subcircuit where the current sources have been removed, as shown in Fig. 7.17(a) on next page. I have chosen J2 as the outer mesh because in that way I have Ix = J1. However, this has been a choice, it does not have to be that way.

Next, sources are reinserted one at a time. Each current source will define at least one loop containing the source and with all other elements not a current source. One possible selection is then the one shown in Fig. 7.17(b), where in addition I have taken care not to have Ix depending on two current loops.

With this selection, we see that J4 = 4 mA is a known value, and J3 = 20 Ix = 20J1 defines an equation. The equations, in expanded form, are then
Loop 1:

$$(250 + 2200 + 1200 + 575) J_1 - (1200 + 575) J_2 + 2200\, J_3 = 6 + (2200 + 1200)(4 \times 10^{-3}) \quad (7.18a)$$

Loop 2

$$-(1200 + 575) J_1 + (1200 + 575 + 250 + 750) J_2 + 750\, J_3 = -1200(4 \times 10^{-3}) \quad (7.18b)$$

Dependent current Source

$$-20\, J_1 + J_3 = \quad (7.18c)$$

The current sources' voltages are not included in these three equations. If any voltage is needed, add the equation of the corresponding loop.

---
[3]This applies equally to symbolic cases, applying the principle of proportionality

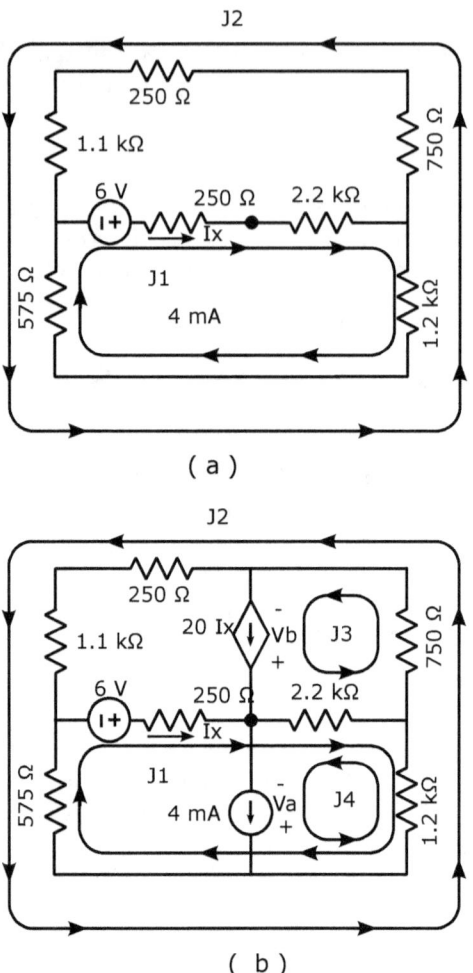

Figure 7.17: Steps in the analysis of circuit in Fig. 7.16: a) step 1: ignore current sources; b) step2 include them, one loop per source

For planar circuits, this method has been called super-mesh method. Again, in my opinion this notation has been included ignoring previous methods of loop selection. One, for example is using trees, and selecting the tree in such a way that no source current is included in it. The loops that result are called fundamental loops, and have been used since long before the term "super mesh" was coined.

7.5 Further considerations

Similar exercises to those done with nodal analysis dealing with superposition, and symbolic and time dependent sources, could be introduced here. You already have the knowledge on how to do it. For the sake of not making this reference book longer, I will skip that part and leave to the interested reader to build up examples. Besides, the next chapters will provide further insight on these applications.

Remember, being able to state and solve the equations is only part of the game. A very important part, because without this skill there is no more way to follow. But what makes the road interesting, is the applications you are able to do with the results. In this sense, I think that the use of the calculator enhances its value as a tool, because not only provide us a tool to solve equations, but with appropriate considerations in the process, we can see interpretations for applications directly from the solution.

Let's continue!

CHAPTER 8

Network Functions

Network functions provide us with information about what we should expect as a response for a given input signal, without necessarily knowing the details of what is happening inside the circuit. By input signal, we refer to a voltage, current, or power present at the port called *input port*. The response may be the voltage or current at an element, or the power absorbed. The element is sometimes called *load*, and the response is usually called *output*.

This chapter focuses on finding the network functions of a circuit. Actually, what we do is to interpret results in the solutions found with the analysis using any of the methods already discussed. Or any other that the reader may prefer. Sometimes additional equations may be introduced in the set for convenience.

8.1 Definition of the network functions

This chapter is focused on circuits with one independent source, which provides the *input signal*, either voltage or current. The two terminals to which the source is connected define the input port. The signal of interest, the *output signal*, may be located at the input port or else it may be a voltage or current at another element called load, which may also be an open circuit or a short-circuit. This situation is illustrated by Fig. 8.1. Notice that the input current I_{in} enters the circuit through the terminal connected to the plus sign of the voltage V_{in}. The input power $P_{in} = V_{in} I_{in}$ is therefore the power generated by the source and absorbed by the circuit.

In general, denote the input and output signals as X_{in} and X_{out}, without any particular reference to the type of magnitude, except if confusion may arise. The output signal could be the current or voltage at the source. In this case, if the input signal is the voltage of the source, then the input current would be the output signal. And the other way around. When both the input and the output signal are at the input port, the network function is called *port function*. Otherwise, it is a *transfer function*.

8.1. DEFINITION OF THE NETWORK FUNCTIONS

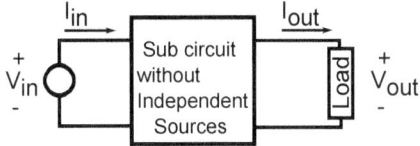

Figure 8.1: One source (input) and load (output) circuit

As far as the process of analysis concerns, the input signal source may be of any type. If it is a voltage source, then v_s is known and i_s is to be calculated, and vice versa. In fact, except for situations where a mathematical inconsistency may arise, the treatment of these magnitudes is a matter of mathematical convenience.

Because of the principle of proportionality discussed in chapter 4, we know that the output signal will be proportional to the input signal. That is, $X_{out} = K X_{in}$, where K is a constant that depends completely on the circuit elements and the topology. In other words, K is defined entirely by the circuit and is independent of the input signal. This allows us to define a *transfer function*, or more accurately, a *network function* with an expression of the form

$$K = \frac{X_{out}}{X_{in}} \tag{8.1}$$

When the input and output signals are defined, then they receive a particular name. The different functions are the following, all making reference to Fig. 8.1:

Equivalent Resistance	$R_{eq} = \dfrac{V_{in}}{I_{in}}$	Equivalent Conductance	$G_{eq} = \dfrac{I_{in}}{V_{in}}$
Current transfer	$A_I = \dfrac{I_{out}}{I_{in}}$	Voltage transfer	$A_v = \dfrac{V_{out}}{V_{in}}$
Trans-resistance	$R_m = \dfrac{V_{out}}{I_{in}}$	Trans conductance	$G_m = \dfrac{I_{out}}{V_{in}}$

In specific applications, these functions are called with other names. For example, what we have called voltage and current transfer functions, in electronics are called voltage and current gains, respectively. When the gain is less then 1, the term attenuation is also used. Similarly, the equivalent resistance may be called input resistance or output resistance, depending on the use or name given to a pair of terminals in an application. These and other notations are specific to a field or application, but refer always to the same definition.

The power transfer function, or power gain, is defined as

$$A_p = \frac{P_{out}}{P_{in}} = \frac{V_{out} I_{out}}{V_{in} I_{in}} \tag{8.2}$$

Again, notice this is the ratio of power absorbed to the power delivered by the source.

8.2 Finding the network functions

Even though finding a network function is just a matter of interpreting results, some students have problems relationg the analysis methods to the goal of calculating a network function.

In an attempt to explain this relationship, let me break the process in two parts. First, laying out the principles. Second, illustrating with examples.

8.2.1 Principles of calculation

You are going to find a network function by proceeding with an analysis of the circuit, by whatever method you prefer. But to simplify your task, spend few minutes doing a planning exercise which will consist of steps 1 and 2 below. Once you have done your work there, proceed to analyze your circuit, and calculate the numerical value following the guidelines of step 3

1. **Placing the source:** Before starting, connect a source between the terminals where your input signal source should go. If you are working a power transfer calculation, you will need both voltage and current of your source.

    (a) If the input signal is described only as a current or voltage at a node, then the other terminal is ground.

    (b) If you plan to work using a reduction/transformation process, it may be easier if you use a source of the same type as your signal.

    (c) When setting up equations (nodal, loop, or other), you may use any type of source, but insure that the input signal type is included. In general, I prefer to include both voltage and current of the source in the equations, one of them to be considered "known". I do it this way because then I have access to all functions. If you don't do that, be sure to do the following:

    - If you use a current source but calculations assume a voltage input signal, include the source voltage as a variable in the equations.
    - Inversely, for a voltage source but current signal, include the current of the source as a variable.
    - If you are to find power transfer, both voltage and current of the source *must* be included in the equations.

2. **Recognize your output signal** Identify your output signal, voltage or current. If it's power, associate it to a voltage, or current, or both. We refer next to voltage or output only.

    (a) If you are using a reduction/transformation method, try not to hide your output signal through the process.

    - If your method hides it, keep track of it by writing up the equations or steps you will follow to recover it.

8.2. FINDING THE NETWORK FUNCTIONS

(b) If you set up equations, like nodal, or loop, or a combination, be sure to associate your signal and your variables

- When possible, choose the method that gives you the output signal directly; that is, a variable in your system of equations is the voltage or current that constitutes the output signal.
- If your output signal is not one of the variables in the system of equations, be sure to write down an extra equation relating the variables and the output signal. This equation may be incorporated in the system, adding one variable –the output signal – or worked out separately.

3. **Find the numerical value of your function:**

   (a) If you used a source of different type as the signal, or else a source with a numeric value different to 1 (one), then divide the output voltage or current by the value of the input signal

   (b) If your source has a numeric value of 1, then the numeric solution may be read directly

Table 8.1 summarizes how to find the numeric value of your function. The symbol $\doteq$ means "numerically equal". Half the values are obtained from the analysis result directly, the other half requires an additional step. Examples are given in the next section.

Table 8.1: Network Functions for unit value sources

Using a 1 A source:			
Equivalent Resistance	$R_{eq} \doteq V_{in}$	Equivalent Conductance	$G_{eq} \doteq \dfrac{1}{V_{in}}$
Current transfer	$A_I \doteq I_{out}$	Voltage transfer	$A_v \doteq \dfrac{V_{out}}{V_{in}}$
Trans-resistance	$R_m \doteq V_{out}$	Trans conductance	$G_m \doteq \dfrac{I_{out}}{V_{in}}$
Using a 1 V source:			
Equivalent Resistance	$R_{eq} \doteq \dfrac{1}{I_{in}}$	Equivalent Conductance	$G_{eq} \doteq I_{in}$
Current transfer	$A_I \doteq \dfrac{I_{out}}{I_{in}}$	Voltage transfer	$A_v \doteq V_{out}$
Transresistance	$R_m \doteq \dfrac{V_{out}}{I_{in}}$	Trans conductance	$G_m \doteq I_{out}$

As a final remark, notice that the functions are not completely independent. Some relations are the following:

$$P_v = A_v \times A_I = R_m \times G_m; \quad R_m = A_v \times R_{eq}; \text{ etc.}$$

However, only the port functions, equivalent resistance and conductance, are inverse one of the other, that is $R_{eq} \times G_{eq} = 1$. For the transfer functions, the inverse **is not** a transfer function. Be aware of that.

8.2.2 Calculating functions: examples

As you might have imagined, the different examples presented for reduction to a single equivalent resistance, are examples for that function, and its inverse, of course. (See examples 5.1 on page 50 and those that follow). Similarly, the voltage and current divider formulas are examples for calculating the voltage and current transfer functions, respectively, for particular cases.

Let us now take other examples, starting with a previously worked circuit. We can thus see what we modify in the process when we focus only in specific results, not all of them.

Example 8.1 *Let us take the circuit of example 5.5 on page 57, with the 20 V source as the input source, and the 605 Ω resistance as the load. For easy reference, let me reproduce the notations in Fig. 8.2. From this figure, we see that I_6 is the input current, and V_4 and I_4 are the output voltage and current, respectively.*

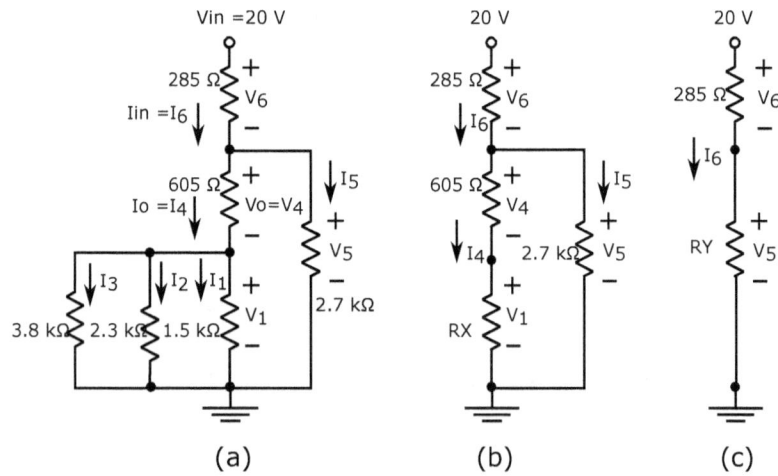

Figure 8.2: Circuit from example 5.5

Let us create the list **L6** for all these variables, with the meaning $L6 = \{ V_{in}, I_{in}, I_{out}, V_{out}\}$. As for the differences with example 5.5, one is that the same steps but up to the point where V_4 and I_4 are available. The second difference is that

8.2. FINDING THE NETWORK FUNCTIONS

we store the value of the input and output variables in L6, *so we can proceed to calculate the functions.*

Initialize signals list L6: {20, 0,0,0} [STO▶] L6 [ENTER]

Create list L1: {1.5 [EE] 3, 2.3 [EE] 3, 3.8 [EE] 3} [STO▶] L1 [ENTER]

Calculate Rx as X: 1/sum(1/L1) [STO▶] X [ENTER] (Rx = 732.81 Ω)

Create list L2: {X + 605, 2.7 [EE] 3} [STO▶] L2 [ENTER]

Calculate Ry: 1/sum(1/L2) [STO▶] Y   (Ry = 897.84 Ω).

<u>**Calculate I_6 and store:**</u> 20 ÷ (285 + Y) [STO▶] L6[2] [ENTER] → 16.909 E-3
(I_6 = 16.91 mA)

Calculate V_5: × Y [ENTER] → 15.181 E0   (V_5 = 15.18 V)

<u>**Calculate I_4 and store:**</u> ÷ (X + 605) [STO▶] L6[3] [ENTER] → 11.348E-3
(I_4 = 11.35 mA)

<u>**Calculate V_4 and store:**</u> × 605 [STO▶] L6[ ]) [ENTER] → 6.865E0
(V_4= 6.87 V)

Check List L6 : L6 [ENTER] → {20.000E0   16.909E-3   11.348E-3   6.865E0 }

Observe that with the exception of V5, no other current or voltage was calculated. Now that we have the list of input and output signals, we are able to find the six network functions as follows:

L6/L6[1] → {1.E0, 845.45E-6, 567E-6, 343.27E-3}

is a result that shows the numerical values of {1, G_{eq}, G_m, A_v}, *respectively.*

L6/L6[2] → {1.1828E3, 1.E0, 671.12E-3, 406.02E-3}

is a result that shows the numerical values of { R_{eq}, 1, A_I, R_m}, *respectively.*

For the power transfer, $A_p = (V_{out} I_{out})/(V_{in} I_{in}$, *do*

L6[3]*L6[4]/(L6[1]*L6[2]) → 230.38E-3

As you can see from this example, the functions have been obtained by *proper interpretation* of the results. Of course, if the original purpose is to find the functions in the first place, than it is wiser to use a 1 V source, and the first set of results are directly obtained. Also, you would stop the process once the output current and voltage is found, without finding the rest of magnitudes.

Before going to the next example, let's make a brief detour to facilitate numbers and readings:

Scaling: If all resistances are provided in kΩ units (conductances in mS units), then the equivalent resistance R_{eq} and trans-resistance R_m functions are in kΩ units too, while the equivalent conductance G_{eq} and trans-conductance G_m functions are in mS units. The voltage and current transfer functions, A_v, A_I, as well as the power transfer A_P remain unaltered.

Example 8.2 *For the circuit of Fig. 8.3, find the input resistance (that is, the equivalent resistance seen at the input) and the voltage gain (that is, A_v).*

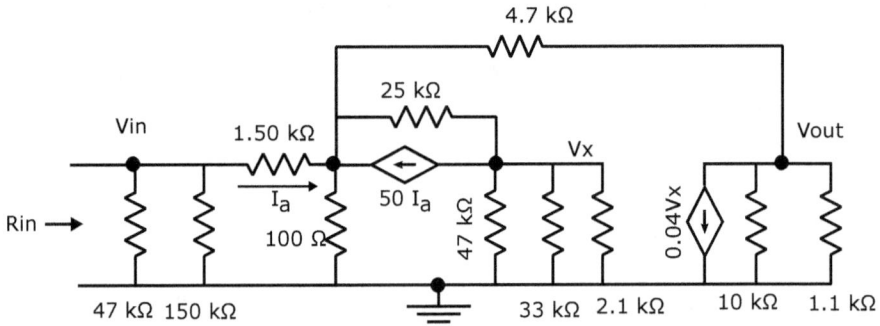

Figure 8.3: An example for Network Functions

Solution with current source: *Let us solve this problem, once using a current source, and the second time using a voltage source. For the current source take Fig. 8.4. The input voltage is V_{N1}, or V_1 for short, and the output voltage is V_4.*

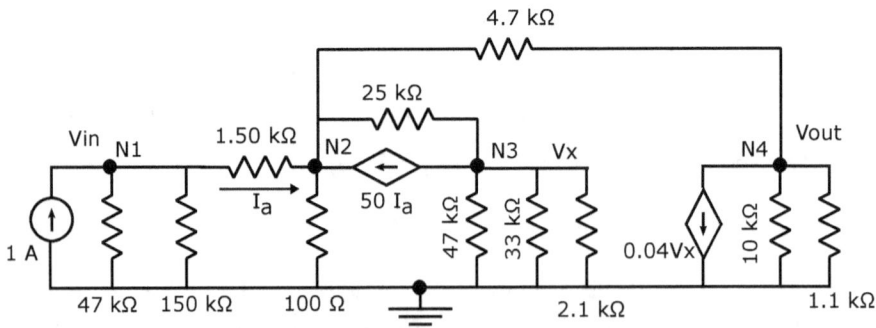

Figure 8.4: Solution using a current source

Using kΩ units for resistances, the nodal admittance matrix for this circuit, following the guidelines of chapter 6 may be obtained with

8.2. FINDING THE NETWORK FUNCTIONS

$$\mathbf{Y} = \begin{array}{c} \\ Node1 \\ Node2 \\ Node3 \\ Node4 \end{array} \begin{bmatrix} V_1 & V_2 & V_3 & V_4 \\ \frac{1}{47}+\frac{1}{150}+\frac{1}{1.5} & -\frac{1}{1.5} & 0 & 0 \\ -\frac{51}{1.5} & \frac{51}{1.5}+\frac{1}{0.1}+\frac{1}{25}+\frac{1}{4.7} & -\frac{1}{25} & -\frac{1}{4.7} \\ \frac{50}{1.5} & -\frac{50}{1.5}-\frac{1}{25} & \frac{1}{25}+\frac{1}{47}+\frac{1}{33}+\frac{1}{2.1} & 0 \\ 0 & -\frac{1}{4.7} & 40 & \frac{1}{4.7}+\frac{1}{1.1}+\frac{1}{10} \end{bmatrix}$$

Notice that the transconductance 0.04 S of the dependent source connected to the output was entered in mS units. that is 40 mS. The known vector is $\mathbf{I_N} = [1, 0, 0, 0]^T$. The solution $\mathbf{Y}^{-1}\mathbf{I_N}$ yields the following result, where I have labeled each row for easy reading:

$$\begin{array}{c} (Vin) \\ (V2) \\ (V3) \\ (Vout) \end{array} \begin{bmatrix} 22.854 \; E0 \\ 22.312 \; E0 \\ -30.252 \; E0 \\ 994.25 \; E0 \end{bmatrix} \quad (8.3)$$

The Vin value is numerically equal to the resistance seen by the current source, in kΩ units. Therefore $R_{in} = 22.854$ kΩ. The voltage transfer function $A_v = V_{out}/Vin = 43.5$ is obtained with

ans(1)[4,1]/ans(1)[1,1] `ENTER` → 43.504

Notice also that $R_m = 994.25$ kΩ, because this is the numerical value of V_{out}. The other functions cannot be determined because no load has been specified. We could think of an open circuit load, or else think of one of the resistances connected to ground and output as the load.

Solution with voltage source: Now let us introduce a voltage source as input, as illustrated in Fig. 8.5. Notice that we have included I_{in} as a variable to be used because this value is required to find the equivalent resistance seen by the source, i. e. , the input resistance.

Again, following the guidelines and methods of chapter 6, the coefficient matrix and known vector are, respectively,

$$\mathbf{Y} = \begin{array}{c} \\ Node1 \\ Node2 \\ Node3 \\ Node4 \end{array} \begin{bmatrix} I_{in} & V_2 & V_3 & V_{out} \\ -1 & -\frac{1}{1.5} & 0 & 0 \\ 0 & \frac{51}{1.5}+\frac{1}{0.1}+\frac{1}{25}+\frac{1}{4.7} & -\frac{1}{25} & -\frac{1}{4.7} \\ 0 & -\frac{50}{1.5}-\frac{1}{25} & \frac{1}{25}+\frac{1}{47}+\frac{1}{33}+\frac{1}{2.1} & 0 \\ 0 & -\frac{1}{4.7} & 40 & \frac{1}{4.7}+\frac{1}{1.1}+\frac{1}{10} \end{bmatrix} \quad (8.4)$$

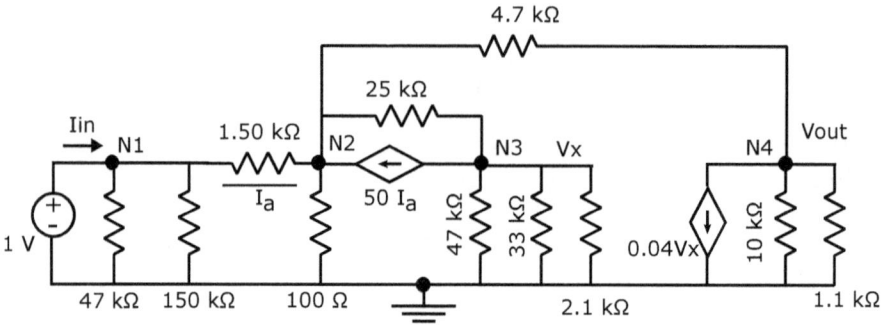

Figure 8.5: Solution using a voltage source

and

$$\mathbf{I_N} = \begin{matrix} Node1 \\ Node2 \\ Node3 \\ Node4 \end{matrix} \begin{bmatrix} -\left(\frac{1}{47} + \frac{1}{150} + \frac{1}{1.5}\right) \\ \frac{51}{1.5} \\ -\frac{50}{1.5} \\ 0 \end{bmatrix} V_{in} \quad (8.5)$$

The solution for the equation, $\mathbf{Y}^{-1}\mathbf{I_N}$, results in

$$\begin{matrix} (Iin) \\ (V2) \\ (V3) \\ (Vout) \end{matrix} \begin{bmatrix} 43.755\text{E-}3 \\ 976.28\text{E-}3 \\ -1.3237\text{E}0 \\ 43.504\text{E}0 \end{bmatrix} \quad (8.6)$$

We see that $V_{out} = 43.504$ V, numerically equal to A_v, being the same result as before. Also, $I_{in} = 43.755$ E-3 mA, which is numerically equal to the equivalent conductance in mS units. Hence, in kΩ units the equivalent resistance is obtained as

ans(1)[1,1]$^{-1}$ ENTER $\rightarrow$ 22.854E0, which is the same result as before.

8.3 Open and short circuit transfer functions

Sometimes it is necessary to calculate network functions under the special circumstances when the load is either an open circuit or a short circuit. Normally, the

8.3. OPEN AND SHORT CIRCUIT TRANSFER FUNCTIONS

open circuit case is not so problematic, as example 8.2 has illustrated, since the output can be considered a voltage output with $I_{out} = 0$ A, which is precisely an open circuit.

Now, when the conditions require a short circuit, there are two forms to consider this problem. One, is to see if the short circuit current belongs to an element too when the short-circuit is introduced, in which case the problem is solved by considering this element to be the load.

The other, is when we require to introduce the short circuit and the current is not contained in an element. We may deal with the problem by applying the Kirchhoff Current equation there. Or else, we can introduce a voltage source at the desired terminals, and then the short circuit occurs when this source becomes 0.

Let us better illustrate this remark with an example.

Example 8.3 *Consider again the circuit of Fig. 8.3, but now we also want to find the output resistance R_{out}, which is the equivalent resistance seen from the output terminals.*

Since we are looking at the circuit from another pair of terminals, where we want to find an equivalent resistance, the problems becomes a candidate for superposition since, by definition, the equivalent resistance requires the subcircuit to be without independent sources.

That means that the source signal must be 0 when calculating the output resistance. There is hence a difference whether we consider the signal to be a current or a voltage. In this case, **the source must be of the same type as the input signal!** *The problem statement is incomplete, since it does not specify which type of input signal should we consider!*

Let us work it for a voltage input signal, and use the circuit of Fig. 8.5. This means in practical terms that R_{out} will be calculated with the input terminals in short circuit. I recommend the reader to use a current source for the open circuit case, and compare the results.

On the other hand, example 8.2 assumed always that the output was an open circuit. Therefore, to calculate the output resistance, we need to introduce a current source as our option. Otherwise, when applying superposition we will short circuit the output to ground! In conclusion, the circuit to be solved is that of Fig. 8.6.

For this circuit, the nodal equations used are in the form $\mathbf{YV} = \mathbf{I_N}$*, where* $\mathbf{Y}$ *is the same as (8.4), while* $\mathbf{I_N}$ *is*

$$\mathbf{I_N} = \begin{matrix} Node1 \\ Node2 \\ Node3 \\ Node4 \end{matrix} \begin{bmatrix} \begin{matrix} V_{in} & I_T \end{matrix} \\ -\left(\frac{1}{47} + \frac{1}{150} + \frac{1}{1.5}\right) & 0 \\ \frac{51}{1.5} & 0 \\ -\frac{50}{1.5} & 0 \\ 0 & 1 \end{bmatrix}$$

Observe that the first column is the vector (8.5). The solution for the system is

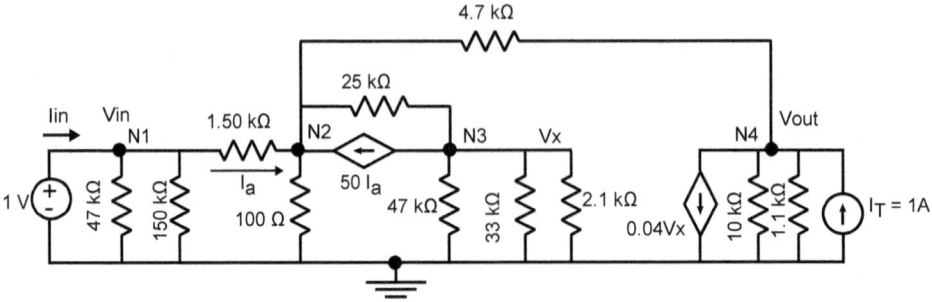

Figure 8.6: For an open circuit transfer function, and a short circuit output resistance calculation

$$\begin{matrix}(Iin)\\(V2)\\(V3)\\(Vout)\end{matrix}\begin{bmatrix} 43.755\text{E-}3 & -257.24\text{E-}6 \\ 976.28\text{E-}3 & 385.86\text{E-}6 \\ -1.3237\text{E}0 & 22.681\text{E-}3 \\ 43.504\text{E}0 & 75.991\text{E-}3 \end{bmatrix} \quad (8.7)$$

The first column, which is what results if $I_T = 0$ A, has already been interpreted in the previous example. the second column corresponds to the case when $V_{in} = 0$ V. Interpreting results, the numerical value of V_{out} will be equal to the resistance seen by the current source, R_{out}, in $k\Omega$ units, when the input is short circuited. Therefore, $R_{out} = 76 \, \Omega$.

In this example the reader should also appreciate the importance of interpretation when scaling has been applied, because of the units involved.

8.4 Chapter summary

There is an intimate relationship between the network functions and the principle of proportionality, and thus of homogeneity. The practical consequence is that finding the network function becomes a straightforward operation of analysis.

The signal concept is also important. Although it is irrelevant what type of signal source is used when all the functions to be calculated have the same input, the question of what kind of source is convenient becomes important when there is the need to calculate another function when excitation goes to another port, since then the original input source must be turned off, raising the question on whether we want a short circuit or an open circuit function. The reader can verify the different results obtained when sources are changed.

CHAPTER 9

Superposition: A powerful tool

The previous chapter applied the principle of proportionality to find the network functions of two terminal sub circuits without independent sources. But circuit theory and applications also include other important definitions and theorems such as Thevenin and Norton equivalents, two-port parameters, multiports, and so on.

This chapter presents superposition as the theoretical principle used to do these new calculations easily, while also also extending some results usually presented in textbooks.

9.1 Thevenin and Network Equivalent Circuits

We start by calculating Thevenin and Norton equivalent circuits. These are two important concepts in applied circuit theory. A first introduction was given before for particular cases.

9.1.1 Basic theory

Consider a linear resistive subcircuit N to which any kind of element may be connected, as shown in Fig. 9.1(a). Thevenin's theorem states that we can substitute the subcircuit by a voltage source in series with a resistance, as shown in (b), while Norton's theorem says that it can be substituted by a current source in parallel with a resistance, as shown in (c). The equivalent subcircuits are called, respectively, Thevenin and Norton equivalent circuits.

The voltage source V_{th} is called Thevenin's voltage. Since it is the voltage that we have in the particular case in which the element connected to the terminals is an open circuit, in which case $i = 0$, it is also called open circuit voltage V_{oc}. This is illustrated in Fig. 9.2(a).

Similarly, the current source I_N is called Norton's current. Since it is the current that occurs in the particular case in which the element connected to the terminals

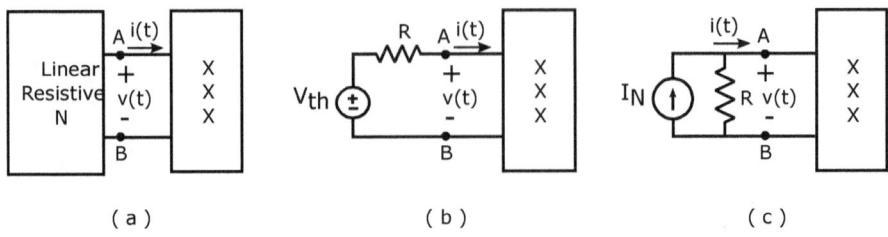

Figure 9.1: (a) General Configuration, (b) Thevenin's equivalent representation (c) Norton's Equivalent Representation

is a short circuit, in which case $v = 0$, it is also called short circuit current I_{sc}. This is illustrated in Fig. 9.2(b).

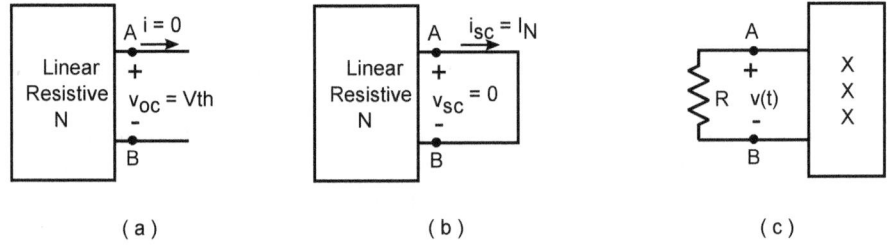

Figure 9.2: (a) Thevenin's voltage. (b) Norton's current (c) Equivalent Resistance R

The resistance R, called Thevenin's resistance or Norton's resistance, depending the representation, is in fact the equivalent resistance seen by the element, which can be calculated by turning off all independent sources in N, as illustrated in Fig. 9.2(c). Notice that Norton's circuit requires $R \neq 0 \, \Omega$ and Thevenin's representation requires $R \neq \infty$. When both representations exist they are in fact related by source transformation, that is,

$$V_{th} = RI_N \tag{9.1}$$

9.1.2 Finding equivalents: principles

If you work a circuit with transformations, you will be able to get an equivalent circuit directly by reducing the original subcircuit to the desired form. Hence, let us look at thos cases in which you write down circuit equations. Let us proceed as follows.

At the terminals at which we want the representation, illustrated by Fig. 9.3(a), substitute the XXX element with a source, that can be either a current source as in Fig. 9.3(b) or a voltage source as in 9.3(c). *Observe the directions of the respective sources.*

9.1. THEVENIN AND NETWORK EQUIVALENT CIRCUITS

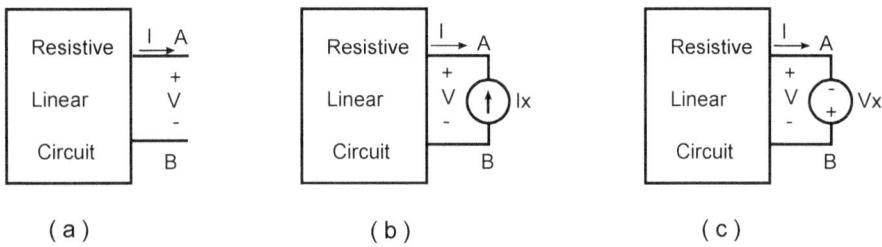

(a) (b) (c)

Figure 9.3: Finding Thevenin's and Norton's equivalent circuits: (a) Target subcircuit (b) using a current source or (b) using a voltage source

Consider the current source connection. By superposition, the voltage at the terminals is of the form $v = a + bI$. Here, a is the voltage when $Ix = 0$, that is, an open circuit, and $v = bIx$ when all other sources – internal to the sub circuit – are off. From the previous chapter, we know then that b is the equivalent resistance seen by the current source. In summary, the voltage v that results from the configuration in Fig. 9.3(b) is

$$v = V_{oc} + RI \tag{9.2}$$

Hence, the two parameters, Thevenin's Voltage and the equivalent resistance R, are directly read from the result if $Ix = 1$ A. Using (9.1) we may now find I_N.

Similarly, the current i in Fig. 9.3(c) is, by superposition, of the form $i = C + dV$, where C is the voltage when $V = 0$, and $i = dV$ when all other sources – internal to the sub circuit – are off. Therefore, taking into account the reference direction of i, the

$$i = I_N + \frac{1}{R}V \tag{9.3}$$

Hence, in one step we have found Norton's current and the equivalent conductance $1/R$, whose inverse is R. With (9.1) we may now find V_{th}.

9.1.3 Finding equivalents: examples

Let us illustrate principles with examples. Let us work the first one using both methods.

Example 9.1 *We are interested in finding the Norton and Thevenin equivalents for the subcircuit of Fig. 9.4 as seen from terminals A and B.*

This will be done using a current source, and a voltage source, separately. The arrangements for the respective cases are shown in Fig. 9.5. The current source case is prepared for nodal analysis, while that with the voltage source for loop analysis. A source transformation was included for the sake of easier procedure.

Solving with the current source: *Since we are not interested on the current flowing in the voltage source, let us use the "super node" equations, joining nodes*

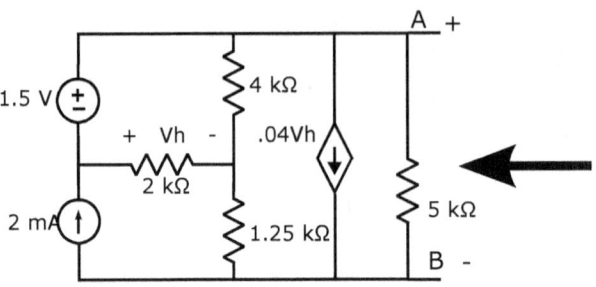

Figure 9.4: Target subcircuit for Thevenin and Norton equivalent circuits

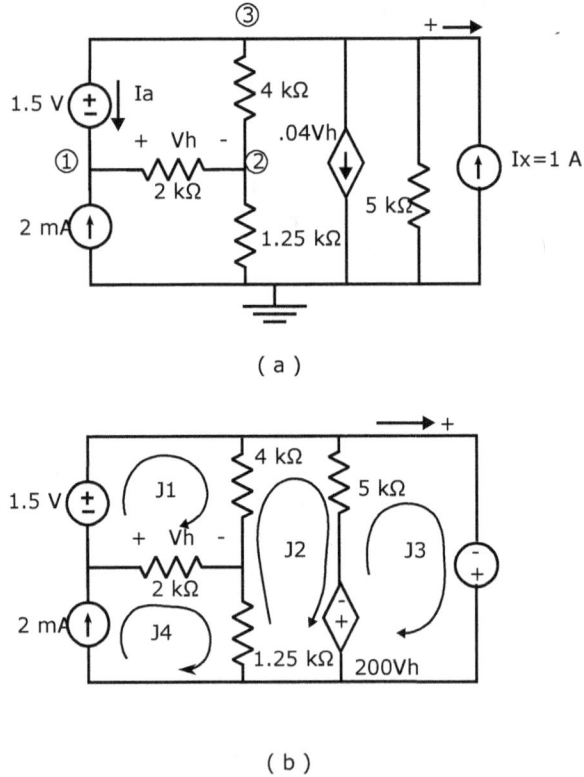

Figure 9.5: (a) Prepared for solving with current source; (b) prepared for voltage source solution

9.1. THEVENIN AND NETWORK EQUIVALENT CIRCUITS

1 and 3 in one equation, as shown in (9.4) below. If you have problems inserting the "0.04", remember that the dependent source is going into node 3, which belongs to the supernode, and thus is negative. Write down the equations by hand first if you have doubts.

$$
\begin{array}{c}
Node(1,3) \\ Node2 \\ V-src
\end{array}
\begin{bmatrix}
0.04+\frac{1}{2E3} & -0.04-\frac{1}{2E3}-\frac{1}{4E3} & \frac{1}{4E3}+\frac{1}{5E3} \\
-\frac{1}{2E3} & \frac{1}{2E3}+\frac{1}{4E3}+\frac{1}{1250} & -\frac{1}{4E3} \\
-1 & 0 & 1
\end{bmatrix}
\begin{bmatrix} v_1 \\ v_2 \\ v_3 \end{bmatrix}
=
\begin{bmatrix} 0.002 & 1 \\ 0 & 0 \\ 1.5 & 0 \end{bmatrix}
\quad (9.4)
$$

with columns labeled V_1, V_2, V_3 and right-hand side columns "In" and Ix.

The solution for this system is, after entering the coefficient matrix as `yn`, and the knowns as `in`,

$$
\mathrm{yn}^{-1} * \mathrm{in} \to
\begin{bmatrix}
526.74\mathrm{E}\text{-}3 & 47.098\mathrm{E}0 \\
496.818\mathrm{E}\text{-}3 & 22.789\mathrm{E}0 \\
2.027\mathrm{E}0 & 47.098\mathrm{E}0
\end{bmatrix}
$$

Since we are looking the subsystem between node 3 and ground, the solution comes out from reading the result for this voltage, at the third row: Vth = 2.027 V and a negative equivalent resistance Rth = 47.098 Ω. The Norton Current source is found using (9.1), working with the values already in the calculator:

`ans(1)[3,1]/ans(1)[3,2]` $\to$ 42.032E-3

which we interpret as a short circuit Norton current IN = 42.032 mA.

Solving with the voltage source: Let us now take the circuit in Fig. 9.5(b). Again, since we are not interested on the voltage at the 2 mA current source, we can write the system as $\mathbf{Z_M}\,\mathbf{J_M} = \mathbf{V_L}$, where $\mathbf{J_M} = \begin{bmatrix} J_1, J_2, J_3 \end{bmatrix}^T$,

$$
\mathbf{Z_M} = \begin{array}{c} Loop1 \\ Loop2 \\ Loop3 \end{array}
\begin{bmatrix}
4E3+2E3 & -4E3 & 0 \\
-4E3+200(2E3) & 4E3+1250+5E3 & -5E3 \\
-200(2E3) & -5E3 & 5E3
\end{bmatrix}
\quad (9.5a)
$$

with columns J_1, J_2, J_3.

and

$$
\mathbf{V_L} = \begin{array}{c} Loop1 \\ Loop2 \\ Loop3 \end{array}
\begin{bmatrix}
0.002(2E3)+1.5 & 0 \\
0.002(1250)+0.002(200)(2E3) & 0 \\
-0.002(200)(2E3) & 1
\end{bmatrix}
\quad (9.5b)
$$

with columns VL and Vx.

The solution for this system is, after entering the coefficient matrix as zm, and the knowns as vl,

$$\mathtt{zm}^{-1} * \mathtt{vl} \rightarrow \begin{bmatrix} 2.508\text{E-}3 & 258.065\text{E-}6 \\ 2.387\text{E-}3 & 387.097\text{E-}6 \\ 43.032\text{E-}3 & 21.232\text{E-}3 \end{bmatrix}$$

Let us store this answer in variable B. Looking at the solution for J3, the first column shows Norton's current IN = 43.032 mA, which is the same as before. The equivalent conductance is 21.232E-3 S whose inverse is the equivalent resistance:

1/B[3,2] → 47.098E0, which corresponds to R_{th} = 47.098 Ω

The Thevenin's voltage is obtained with the relationship $V_{th} = I_N\, R_{th} = I_N/G_{eq}$:

B[3,1]/B[3,2] → 2.027E0, which coincides with the previous result Vth = 2.027 V

The example shows that the analysis method or preferred source does not determine the validity of the process. It must be emphasized that we are assuming that the system has a solution, since there are situations in which this does not happen. For example, in those cases in which a voltage source is in parallel with the port (R=0) or a current source in series with the port (R = ∞).

Exercise suggested: to solve the problem now by generating the coefficient matrices with the programs for nodal and mesh analyses introduced in previous chapters.

When the circuit contains symbolic sources, then the Thevenin Voltage will also be of symbolic type. It can also be obtained as a function of several sources. The following example illustrates this.

Example 9.2 *Assume that the independent sources in example 9.1 are symbolic. That is, in Fig. 9.4 assume that instead of the 1.5 V and 2 mA sources we have a voltage source E_a and a current source I_b. We can find the equivalent representations using the methods that have been used for dealing with symbolic sources. Let us work it next.*

Solving with the current source: The coefficient matrix in 9.4 will not change, but the "known" matrix I_N becomes

$$\mathbf{I_N} = \begin{matrix} Node(1,3) \\ Node2 \\ V-src \end{matrix} \begin{matrix} Va & Ib & Ix \\ \begin{bmatrix} 0 & 1 & 1 \\ 0 & 0 & 0 \\ 1 & 0 & 0 \end{bmatrix} \end{matrix} \qquad (9.6)$$

The solution for this system is, after entering again the coefficient matrix as yn –which you might already have –, and the knowns as in,

9.2. CALCULATING TWO-PORT PARAMETERS

$$yn^{-1} * in \to \begin{bmatrix} 288.4E\text{-}3 & 47.1E0 & 47.098E0 \\ 300.8E\text{-}3 & 22.8E0 & 22.789E0 \\ 1.288E0 & 47.1E0 & 47.098E0 \end{bmatrix}$$

Looking the solution at the third row, we interpret the open circuit voltage as a linear combination of the sources: Vth = 1.288 Va + 47.1 Ib. The equivalent resistance Rth = 47.1 Ω appears here as expected. The Norton Current source is found using (9.1), working with the values already in the calculator:

ans(1)[3]/ans(1)[3,3] → [27.355E-3 1.E0 1.E0]

which we interpret as IN = 27.355E-3 Va + Ib. The equivalent representations for these results are shown in Fig. 9.6.

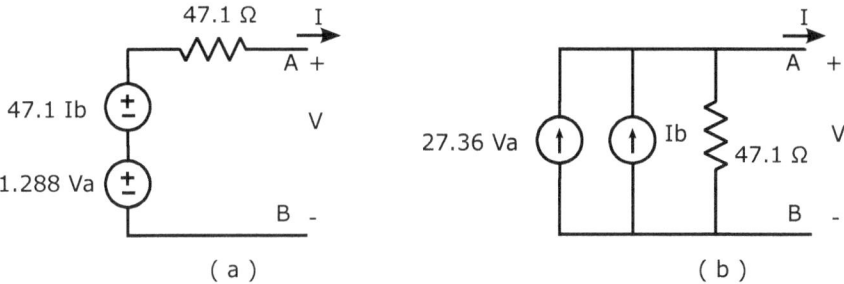

Figure 9.6: (a) Thevenin Equivalent for this example; (b) Norton Equivalent for this example

As you might have imagined from this example, it is possible to extend the applications of superposition to calculate another variations such as for example when one of the sources is dependent on a magnitude which is not part of the subcircuit itself. We leave that sort of discussion outside from this book.

Further applications of the Thevenin and Norton's theorems and calculations will be left for illustrations in Chapter ??. Let us for the moment look at another topic.

9.2 Calculating two-port parameters

The topic of two-port is important enough to deserve its own chapter. Here I limit myself to present the application of superposition to find the parameters. I introduce also an extension of the concept for extended Thevenin and Norton equivalents.

If the reader is not familiar with two-ports yet, a brief introduction is found in the next chapter. Also, the reader may consult circuit textbooks.

9.3 Finding two-port parameters from equations

Two-port parameters may be calculated in different ways:

a) By direct application of definitions: Once we short-circuit or open-circuit terminals, we apply the methods studied in previous chapters. This is the usual way in which they are taught in textbooks. And important also.

b) Obtaining another set of parameters and using conversion formulas. Tables of translation formulas are usually available from textbooks, and also discussed in the next chapter.

c) Connecting sources to both ports and solving the equations for dependent port variables in terms of the independent variables. This is the method to be discussed next. The parameters result from coefficient comparison of results

I illustrate the method using one circuit and deriving two sets of parameters. The equations must include both current and voltages of the added sources.

Example 9.3 *Find the h parameters of the two port shown in Fig. 9.7.*

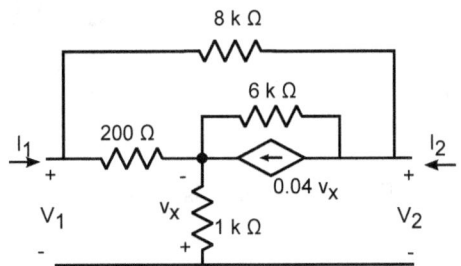

Figure 9.7: Two-port example

I will work this example using mesh equations, with a previous source transformation, as shown in Fig. 9.8. Notice that I do not show the type of source being used. This is irrelevant for our purposes since the method is based on variable manipulation.

Before we start working, let us consider what we are looking for, in order to see how to write down the equations. Since we want h-parameters, that means that our goal is to arrive at equations of the form

$$V_1 = h_{11} I_1 + h_{12} V_2$$
$$I_2 = h_{21} I_1 + h_{22} V_2$$

Hence, V_1 and I_2 are supposed to be unknown, together with I_3, while I_1 and V_2 are considered "symbolic" known sources. Using the methods for loop analysis that we have already studied for such a case, **not forgetting to include the voltage at port 1 as variable**, we arrive at

9.3. FINDING TWO-PORT PARAMETERS FROM EQUATIONS

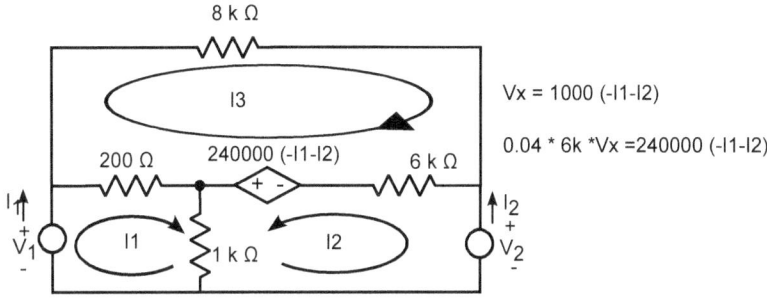

Figure 9.8: Two-port example arranged for loop equations

$$\begin{array}{c} & \begin{array}{ccc} V_1 & I_2 & I_3 \end{array} \\ \begin{array}{c} Loop1 \\ Loop2 \\ Loop3 \end{array} & \left[\begin{array}{ccc} -1 & 1000 & -200 \\ 0 & 247000 & 6000 \\ 0 & 246000 & 14200 \end{array} \right] \end{array} \left[\begin{array}{c} V_1 \\ I_2 \\ I_3 \end{array} \right] = \begin{array}{c} \begin{array}{cc} I_1 & V_2 \end{array} \\ \left[\begin{array}{cc} -1200 & 0 \\ -241000 & 1 \\ -239800 & 0 \end{array} \right] \end{array} \quad (9.7)$$

The solution for this system is

$$\left[\begin{array}{c} V_1 \\ I_2 \\ I_3 \end{array} \right] = \begin{array}{c} \begin{array}{cc} I_1 & V_2 \end{array} \\ \left[\begin{array}{cc} 210.1747 & 00.03121 \\ -0.9764 & 7 \times 10^{-6} \\ 0.02727 & -0.000121 \end{array} \right] \end{array}$$

The H-matrix is formed with the rows corresponding to V_1 and I_2. In this case, the first two rows in that order. Hence

$$\mathbf{H} = \left[\begin{array}{cc} 210.1747 & 00.03121 \\ -0.9764 & 7 \times 10^{-6} \end{array} \right]$$

The previous example has shown the method for calculation directly from equations. Now, what do you want your programmable calculator for if you are always going to write the equations by yourself?

Nah! Let us take advantage of the programs we have written! This become specially true if the circuit is big enough to make you feel unsecure. However, by doing so, *we must be sure to establish first our strategy to work the problem.* Otherwise, we will be lost.

Example 9.4 *The circuit in Fig. 9.9(a) is similar to that of 6.9 on page 96, except that one of the sources has been changed to a dependent source, and two-ports have been identified. For this two port, let us find the ABCD parameters, using the nodal analysis with the program developed in section §6.3.3.*

Before proceeding to use the calculator, let us set up the strategy (plan) we must follow:

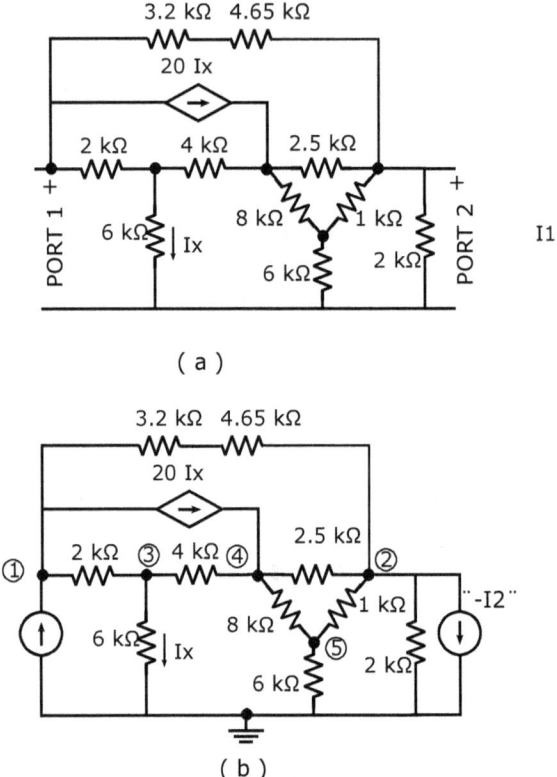

Figure 9.9: Two-port example using nodal program: (a) Two-port definition; (b) Prepared for analysis

9.3. FINDING TWO-PORT PARAMETERS FROM EQUATIONS

Objective To find the ABCD matrix, which correspond to the equations

$$\begin{bmatrix} V_1 \\ I_1 \end{bmatrix} = \begin{bmatrix} A & B \\ C & D \end{bmatrix} \begin{bmatrix} V_2 \\ -I_2 \end{bmatrix}$$

Strategy Use the program **nodal1** to generate the nodal matrix $\mathbf{Y}n$, to set up the equations

$$\mathbf{Y}n\mathbf{V}n = \mathbf{B}$$

where $\mathbf{B}$ is the matrix defined for our circuit with two sources as illustrated in Fig. 9.9(b). Notice that the source at port 2 is already in the desired direction for the chain parameters, and has been labeled with the sign included.

Procedure Rearrange the equations so that the independent variables are V_2 and $-I_2$. (See section §3.5.3). That is, from

$$\mathbf{Y}_{n1}V_1 + \mathbf{Y}_{n2}V_2 + \mathbf{Y}_{n3}V_3 + \mathbf{Y}_{n4}V_4 + \mathbf{Y}_{n5}V_5 = \mathbf{B}_1 I_1 + \mathbf{B}_2(-I_2)$$

we should get

$$\mathbf{Y}_{n1}V_1 + (-\mathbf{B}_1)I_1 + \mathbf{Y}_{n3}V_3 + \mathbf{Y}_{n4}V_4 + \mathbf{Y}_{n5}V_5 = (-\mathbf{Y}_{n2})V_2 + \mathbf{B}_2(-I_2)$$

When solving, our matris is the submatrix formed with the rows for V_1 and I_1.

Having established the logistics, let us now proceed with the calculator.

Step 1: First let us describe our circuit. Define matrices R and G for the resistances and controlled source. The independent current matrix is 0:

$$\mathbf{R} = \begin{bmatrix} 2000 & 1 & 3 \\ 3200+4650 & 1 & 2 \\ 6000 & 3 & 6 \\ 4000 & 3 & 4 \\ 8000 & 4 & 5 \\ 6000 & 5 & 6 \\ 2500 & 4 & 2 \\ 1000 & 2 & 5 \\ 2000 & 2 & 6 \end{bmatrix} \quad G = \begin{bmatrix} 20/6E3 & 1 & 4 & 3 & 6 \end{bmatrix}$$

Now, use the program :nodal1(5,r,0,g) to get

$$\mathbf{yn} = \begin{bmatrix} 627.4E-6 & -127.4E-6 & 2.833E-3 & 0.0 & 0.0 \\ -127.4E-6 & 2.027E-3 & 0.0 & -400E-6 & -1.E-3 \\ -500.E-6 & 0.0 & 916.67E-6 & -250.E-6 & 0.0 \\ 0.0 & -400E-6 & -3.583E-3 & 775.E-6 & -125E-6 \\ 0.0 & -1.E-3 & 0.0 & -125.E-6 & 1.292E-3 \end{bmatrix}$$

Now create the matrix

$$\mathbf{B} = \begin{bmatrix} 1 & 0 \\ 0 & -1 \\ 0 & 0 \\ 0 & 0 \\ 0 & 0 \end{bmatrix}$$

Observe that current I2 is already leaving node 2. To exchange columns, we work by transposing and retrieving and exchanging rows with the respective sign changes. The interested reader may consult section 3.5.3. We use several commands in one entry, taking advantage of the colon (:) notation,

Step 1: yn^T [STO▶] yn : b^T [STO▶] b augment(yn,b)^T [STO▶] z

Meaning: Columns become rows.

Step 2: yn[2] [STO▶] z : [(-)] b[1] [STO▶] yn[2] : [(-)] z [STO▶] b[1]

Meaning: $\mathbf{Y}_{n1}V_1 - \mathbf{B}_1 I_1 + \mathbf{Y}_{n3}V_3 + \mathbf{Y}_{n4}V_4 + \mathbf{Y}_{n5}V_5 = -\mathbf{Y}_{n2}V_2 + \mathbf{B}_2(-I_2)$
still in transposed forn

Step 3: yn^T [STO▶] yn : b^T [STO▶] b augment(yn,b)^T [STO▶] z

Meaning: System back to normal form

Step 4: yn^{-1} * b [STO▶] z

Meaning: Equations solved. Parameter given by the first two rows of z

Step: subMat(z,1,1,2,2)

Meaning: Our set of parameters, solution for our problem.

Following the above steps, the solution to our problem is

$$\begin{bmatrix} A & B \\ C & D \end{bmatrix} = \begin{bmatrix} -546.205E-3 & -227.168E0 \\ 743.358E-6 & 1.1069E0 \end{bmatrix}$$

9.4 Summary

This chapter has shown two applications of superposition using matrices. The mechanism follows in the different applications the same basic principles. Namely, write down the equations in a convenient form so that the solution for the problem you're trying to solve comes out directly, via of course the proper interpretation.

The application of the superposition theorem was applied just to two sort of problems. First, to find the Thevenin and Norton equivalent representations. And second, to calculate the parameters of two-port sub circuits. It should be obvious by now that there are many more possible cases where this method is handy.

CHAPTER 10

Two-port and three-terminal networks

Chapter 9 included a section on how to find the two port parameters using the property of superposition. In this chapter we focus on some theory and applications of two-port network theory, and leave more in depth study for the reader to follow. An interesting feature is that two-ports are attractive for quick calculations, because most applications are mainly based on formulas that can be programmed.

10.1 Basic Definitions

Any device or sub circuit with two or more terminals available for connection is a *multiterminal network*, or *multipole*. If one terminal is taken as a datum or ground with respect to which the potentials of other N terminals are referenced, it is called (N+1)-terminal network, or (N+1)-pole. If the reference node is not part of the network, then it is a *floating* N-pole or floating multipole.

On the other hand, two terminals form a *port* if the current entering one terminal is always equal to the current leaving the other. A 2N terminal network is an *N-port network*, or simply N-port, when the terminals can be associated by pairs to form ports.

Fig. 10.1 shows the general representation of two-ports and three terminal sub circuits. A three pole may be considered a special case of two-port with one pole common to both ports. when it comes to applications. However, although the differences may be important, this topic falls beyond our scope. We focus here only in common properties.

Ports 1 and 2 are often called *input port* and *output port*, respectively, and this convention is used in naming parameters. The reader should however be aware that these names have real meaning only in specific applications.

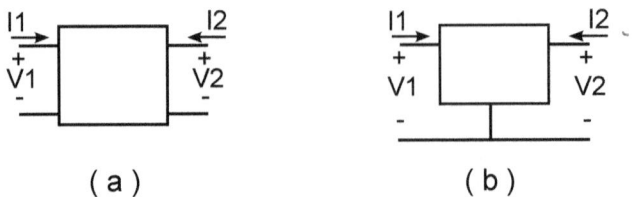

Figure 10.1: (a) Two port and (b) three-terminal networks

10.2 Two port parameters

Two linear equations will characterize the two-port. With four variables involved, V1, V2, I1 and I2, two of them must be selected as independent variables. Six different choices are possible yielding different parameters to characterize the two-port. Some selections may not be possible for particular two-ports.

The definition of the parameters assume that the two-port circuit does not contain any independent source. To find the parameters, therefore, it may be necessary to first turn those sources off. We describe next the different representations.

10.2.1 Definition of Parameters

The six sets of two-port parameters are defined in table 10.1 on the next page. The notations OC, for open-circuit, and SC for short-circuit, arise from the conditions set for the parameter definition. When a current is zero, it becomes an open circuit, and when it is a zero voltage, it becomes short-circuit.

Notice the negative sign for the independent current in the chain parameters. This negative sign stems from a change in the reference direction for the respective current. We shall see the advantage later.

The chain parameters are also knon as ABDC parameters because originally the equations were written, in matrix form, as

$$\begin{bmatrix} V_1 \\ I_1 \end{bmatrix} = \begin{bmatrix} A & B \\ C & D \end{bmatrix} \begin{bmatrix} V_2 \\ -I_2 \end{bmatrix}$$

Similarly, for the b-parameters, the name A'B'C'D',

$$\begin{bmatrix} V_2 \\ I_2 \end{bmatrix} = \begin{bmatrix} A' & B' \\ C' & D' \end{bmatrix} \begin{bmatrix} V_1 \\ -I_1 \end{bmatrix}$$

10.2.2 Transforming parameters

With the exception of chain parameters, there may be sets of parameters that do not exist for one particular two-port network. This section works parameter conversion assuming the target set exists. Two main algorithms can be considered: work conversion as a problem of equations solving, or build up your set of user defined functions. Let us look at both cases.

10.2. TWO PORT PARAMETERS

Table 10.1: Two Port Parameters

Open Circuit impedances or Z-parameters

Matrix equation:

$$\begin{bmatrix} V_1 \\ V_2 \end{bmatrix} = \begin{bmatrix} z_{11} & z_{12} \\ z_{21} & z_{22} \end{bmatrix} \begin{bmatrix} I_1 \\ I_2 \end{bmatrix}$$

Parameter definitions:

OC. Input impedance: $z_{11} = \frac{V_1}{I_1}\Big|_{I_2=0}$

OC. reverse transimpedance: $z_{12} = \frac{V_1}{I_2}\Big|_{I_1=0}$

OC. forward transimpedance: $z_{21} = \frac{V_2}{I_1}\Big|_{I_2=0}$

OC. output impedance: $z_{22} = \frac{V_2}{I_2}\Big|_{I_1=0}$

Short Circuit admittances or Y-parameters

Matrix equation:

$$\begin{bmatrix} I_1 \\ I_2 \end{bmatrix} = \begin{bmatrix} y_{11} & y_{12} \\ y_{21} & y_{22} \end{bmatrix} \begin{bmatrix} V_1 \\ V_2 \end{bmatrix}$$

Parameter definitions:

SC. Input admittance: $y_{11} = \frac{I_1}{V_1}\Big|_{V_2=0}$

SC. Reverse transadmittance: $y_{12} = \frac{I_1}{V_2}\Big|_{V_1=0}$

SC. forward transadmittance: $y_{21} = \frac{I_2}{V_1}\Big|_{V_2=0}$

SC. output admittance: $y_{22} = \frac{I_2}{V_2}\Big|_{V_1=0}$

Hybrid H-Parameters

Matrix equation:

$$\begin{bmatrix} V_1 \\ I_2 \end{bmatrix} = \begin{bmatrix} h_{11} & h_{12} \\ h_{21} & h_{22} \end{bmatrix} \begin{bmatrix} I_1 \\ V_2 \end{bmatrix}$$

Parameter definitions:

SC input impedance: $h_{11} = \frac{V_1}{I_1}\Big|_{V_2=0}$

OC reverse voltage gain: $h_{12} = \frac{V_1}{V_2}\Big|_{I_1=0}$

SC current gain: $h_{21} = \frac{I_2}{I_1}\Big|_{V_2=0}$

OC admittance: $h_{22} = \frac{I_2}{V_2}\Big|_{I_1=0}$

Inverse hybrid G-Parameters

Matrix equation:

$$\begin{bmatrix} I_1 \\ V_2 \end{bmatrix} = \begin{bmatrix} g_{11} & g_{12} \\ g_{21} & g_{22} \end{bmatrix} \begin{bmatrix} V_1 \\ I_2 \end{bmatrix}$$

Parameter definitions:

OC input admittance: $g_{11} = \frac{I_1}{V_1}\Big|_{I_2=0}$

SC reverse current gain: $g_{12} = \frac{I_1}{I_2}\Big|_{V_1=0}$

OC voltage gain: $g_{21} = \frac{V_2}{V_1}\Big|_{I_2=0}$

SC output impedance: $g_{22} = \frac{V_2}{I_2}\Big|_{V_1=0}$

Chain a parameters or ABCD parameters

Matrix equation:

$$\begin{bmatrix} V_1 \\ I_1 \end{bmatrix} = \begin{bmatrix} a_{11} & a_{12} \\ a_{21} & a_{22} \end{bmatrix} \begin{bmatrix} V_2 \\ -I_2 \end{bmatrix}$$

Parameter definitions:

OC voltage gain: $\frac{1}{a_{11}} = \frac{V_2}{V_1}\Big|_{I_2=0}$

SC transadmittance: $\frac{1}{a_{12}} = \frac{-I_2}{V_1}\Big|_{V_2=0}$

OC transimpedance: $\frac{1}{a_{21}} = \frac{V_2}{I_1}\Big|_{I_2=0}$

SC current gain: $\frac{1}{a_{22}} = \frac{-I_2}{I_1}\Big|_{V_2=0}$

Inverse chain b parameters or A'B'C'D' parameters

Matrix equation:

$$\begin{bmatrix} V_2 \\ I_2 \end{bmatrix} = \begin{bmatrix} b_{11} & b_{12} \\ b_{21} & b_{22} \end{bmatrix} \begin{bmatrix} V_1 \\ -I_1 \end{bmatrix}$$

Parameter definitions:

OC reverse voltage gain: $\frac{1}{b_{11}} = \frac{V_1}{V_2}\Big|_{I_1=0}$

SC reverse transadmittance: $\frac{1}{b_{12}} = \frac{-I_1}{V_2}\Big|_{V_1=0}$

OC reverse transimpedance: $\frac{1}{b_{21}} = \frac{V_1}{I_2}\Big|_{I_1=0}$

SC reverse current gain: $\frac{1}{b_{22}} = \frac{-I_1}{I_2}\Big|_{V_1=0}$

Simple cases

The Z and Y matrices are inverse one from the other, and the same goes for the H and G matrices. That is,

$$\mathbf{Z}^{-1} = \mathbf{Y}; \quad \mathbf{Y}^{-1} = \mathbf{Z} \tag{10.1a}$$

$$\mathbf{H}^{-1} = \mathbf{G}; \quad \mathbf{G}^{-1} = \mathbf{H} \tag{10.1b}$$

Hence, these conversions become trivial and will not be further considered.

The chain parameters are not inverse because of the negative signs for the currents, but these can be worked with the matrix element operation dot-star of the TI-89. The conversion is done as

$$\begin{bmatrix} 1 & -1 \\ -1 & 1 \end{bmatrix} . \times \mathbf{A}^{-1} = \mathbf{B}; \quad \begin{bmatrix} 1 & -1 \\ -1 & 1 \end{bmatrix} . \times \mathbf{B}^{-1} = \mathbf{A} \tag{10.1c}$$

Although the four transformations (10.1a) and (10.1b) are really not worth programming, the two shown in (10.1c) may be programmed as functions in the TI-89 taking advantage of the dot operations:

A to B transformation: Define a2b(a)= [1, (-) 1 ; (-) 1 , 1] .* a ^ (-) 1

B to A transformation: Define b2a(b)= [1, (-) 1 ; (-) 1 , 1] .* b ^ (-) 1

Conversion by setting up and solving the equations

The basic principle for conversion by this method is to use the original form and build up the equations $\mathbf{A}x = \mathbf{B}y$, where x is the vector of the new dependent variables, and y that of the new independent set. The 2x2 matrices $\mathbf{A}$ and $\mathbf{B}$ are built by rearrangement of coefficients. This comment is better understood with an example.

Example 10.1 *Let us start from a given y-parameter matrix* $\mathbf{Y}$.

$$\mathbf{Y} = \begin{bmatrix} 232.4E-3 & -12.5E-3 \\ -3.5E-3 & 121.0E-3 \end{bmatrix}$$

To understand the logic of the process, let us look at what the parameters equations mean. We are given the set

$$I_1 = y_{11}V_1 + y_{12}V_2$$
$$I_2 = y_{21}V_1 + y_{22}V_2$$

and we want to arrive at the form

10.2. TWO PORT PARAMETERS

$$V_1 = h_{11}I_1 + h_{12}V_2$$
$$I_2 = h_{21}I_1 + h_{22}V_2$$

Hence, we rearrange the original set of equations that we want to convert so that the dependent and independent variables are identified. That is

$$-y_{11}V_1 = -I_1 + y_{12}V_2$$
$$y_{21}V_1 - I_2 = 0I_1 + y_{22}V_2$$

In matrix form,

$$\begin{bmatrix} -y_{11} & 0 \\ -y_{21} & 1 \end{bmatrix} \begin{bmatrix} V_1 \\ I_2 \end{bmatrix} = \begin{bmatrix} -1 & y_{12} \\ 0 & y_{22} \end{bmatrix} \begin{bmatrix} I_1 \\ V_2 \end{bmatrix}$$

and then solve. The following sequence will generate the h-parameter matrix **H** that results after conversion.

```
[ [ (-) y[1,1],0] [ (-) y[2,1], 1]] STO▶  x ENTER
[ [ (-) 1,y[1,2]] [0,y[2,2]]]  STO▶  B ENTER
x⁻¹  ×  B STO▶  h ENTER
```

Following the above steps results in the target matrix

$$\mathbf{H} = \begin{bmatrix} 4.303 & 0.067 \\ -0.015 & 0.121 \end{bmatrix}$$

Programming transformations

Tables 10.2, 10.3 and 10.4 shows formulas for transformations. From these tables, you may define different functions and have your data base for transformations in your calculator.

To illustrate the remark, from the table it is straightforward to go to the function as illustrated by the following Y to H function definition:

```
Define y2h(y) = (1 ÷ y[1,1])*[[1, (-) y[1,2]] [y[2,1], det(y)]]
```
(10.2)

You may check the formulas yourself. For example, I will illustrate here the Y to H transformation. Start with the Y equations and solve for the dependent magnitudes of the H system:

$$\begin{aligned} y_{11}V_1 + y_{12}V_2 &= I_1 \\ y_{21}V_1 + y_{22}V_2 &= I_2 \end{aligned} \Rightarrow \begin{aligned} V_1 &= \frac{1}{y_{11}}I_1 + \frac{-y_{12}}{y_{11}}V_2 \\ I_2 &= \frac{y_{21}}{y_{11}}I_1 + \frac{y_{11}y_{22} - y_{12}y_{21}}{y_{11}}V_2 \end{aligned}$$

Therefore, by comparison

$$h_{11} = \frac{1}{y_{11}}; \quad h_{12} = \frac{-y_{12}}{y_{11}}; \quad h_{21} = \frac{y_{21}}{y_{11}}; \quad \text{and} \quad h_{22} = \frac{\det(y)}{y_{11}} \qquad (10.3)$$

where $\det(y) = y_{11}y_{22} - y_{12}y_{21}$. These are the formulas found in the table.

Table 10.2: Two-port parameters Z and Y conversion formulas

$$z_{11} = \frac{\Delta h}{h_{22}}; \quad z_{12} = \frac{h_{12}}{h_{22}} \qquad y_{11} = \frac{1}{h_{11}} \quad y_{12} = -\frac{h_{12}}{h_{11}}$$

$$z_{21} = -\frac{h_{21}}{h_{22}}; \quad z_{22} = \frac{1}{h_{22}} \qquad y_{21} = \frac{h_{21}}{h_{11}} \quad y_{22} = -\frac{\Delta h}{h_{11}}$$

$$z_{11} = \frac{1}{g_{11}}; \quad z_{12} = -\frac{g_{12}}{g_{11}} \qquad y_{11} = \frac{\Delta g}{g_{22}} \quad y_{12} = \frac{g_{12}}{g_{22}}$$

$$z_{21} = \frac{g_{21}}{g_{11}}; \quad z_{22} = \frac{\Delta g}{g_{11}} \qquad y_{21} = -\frac{g_{21}}{g_{22}} \quad y_{22} = -\frac{1}{g_{22}}$$

$$z_{11} = -\frac{a_{11}}{a_{21}}; \quad z_{12} = \frac{\Delta a}{a_{21}} \qquad y_{11} = \frac{a_{22}}{a_{12}}; \quad y_{12} = -\frac{\Delta a}{a_{12}}$$

$$z_{21} = \frac{1}{a_{21}}; \quad z_{22} = -\frac{a_{22}}{a_{21}} \qquad y_{21} = 1\frac{1}{a_{12}}; \quad y_{22} = \frac{a_{11}}{a_{12}}$$

$$z_{11} = \frac{b_{22}}{b_{21}}; \quad z_{12} = \frac{1}{b_{21}} \qquad y_{11} = \frac{b_{11}}{b_{12}}; \quad y_{12} = -\frac{1}{b_{12}}$$

$$z_{21} = \frac{\Delta b}{b_{21}}; \quad z_{22} = \frac{b_{11}}{b_{21}} \qquad y_{21} = -\frac{\Delta b}{b_{12}}; \quad y_{22} = \frac{b_{22}}{b_{12}}$$

You may work the algebra for other cases or else consult tables of transformations. Observe that all transformations involve dividing either by a determinant or by one parameter. When the division is by zero, the calculator will warn about the error. This means that the target set of parameters does not exist. Remember, however, that due to rounding errors, sometimes the division is not by zero but by a very small number, yielding unreasonable results. Exercise your judgement!

10.2. TWO PORT PARAMETERS

Table 10.3: Two-port parameters H and G conversion formulas

$h_{11} = \dfrac{\Delta z}{z_{22}}; \quad h_{12} = \dfrac{z_{12}}{z_{22}}$	$g_{11} = \dfrac{1}{z_{11}} \quad g_{12} = -\dfrac{z_{12}}{z_{11}}$
$h_{21} = -\dfrac{z_{21}}{z_{22}}; \quad h_{22} = \dfrac{1}{z_{22}}$	$g_{21} = \dfrac{h_{21}}{z_{11}} \quad g_{22} = -\dfrac{\Delta z}{z_{11}}$
$h_{11} = \dfrac{1}{y_{11}}; \quad h_{12} = -\dfrac{y_{12}}{y_{11}}$	$g_{11} = \dfrac{\Delta y}{y_{22}} \quad g_{12} = \dfrac{y_{12}}{y_{22}}$
$h_{21} = \dfrac{y_{21}}{y_{11}}; \quad h_{22} = \dfrac{\Delta y}{y_{11}}$	$g_{21} = -\dfrac{y_{21}}{y_{22}} \quad g_{22} = -\dfrac{1}{y_{22}}$
$h_{11} = \dfrac{a_{12}}{a_{22}} \quad h_{12} = \dfrac{\Delta a}{a_{22}}$	$g_{11} = \dfrac{a_{21}}{a_{11}} \quad g_{12} = -\dfrac{\Delta a}{a_{11}}$
$h_{21} = \dfrac{1}{a_{22}} \quad h_{22} = -\dfrac{a_{21}}{a_{22}}$	$g_{21} = \dfrac{1}{a_{21}} \quad g_{11} = \dfrac{a_{12}}{a_{11}}$
$h_{11} = \dfrac{b_{12}}{b_{11}} \quad h_{12} = \dfrac{1}{b_{11}}$	$g_{11} = \dfrac{b_{21}}{b_{22}} \quad g_{12} = -\dfrac{1}{b_{22}}$
$h_{21} = \dfrac{\Delta b}{b_{11}} \quad h_{22} = \dfrac{b_{21}}{b_{11}}$	$g_{21} = \dfrac{\Delta b}{b_{22}} \quad g_{22} = \dfrac{b_{12}}{b_{22}}$

Table 10.4: Two-port chain parameters conversion formulas

$$a_{11} = \frac{z_{11}}{z_{21}} \quad a_{12} = \frac{\Delta z}{z_{21}}$$

$$a_{21} = \frac{1}{z_{21}} \quad a_{22} = \frac{z_{22}}{z_{21}}$$

$$b_{11} = \frac{z_{22}}{z_{12}} \quad b_{12} = \frac{\Delta z}{z_{12}}$$

$$b_{21} = \frac{1}{y_{12}} \quad b_{22} = -\frac{z_{11}}{y_{12}}$$

$$a_{11} = -\frac{y_{22}}{y_{21}} \quad a_{12} = -\frac{1}{y_{21}}$$

$$a_{21} = -\frac{\Delta y}{y_{21}} \quad a_{22} = -\frac{y_{11}}{y_{21}}$$

$$b_{11} = -\frac{y_{11}}{y_{12}} \quad b_{12} = -\frac{1}{y_{12}}$$

$$b_{21} = -\frac{\Delta y}{y_{12}} \quad b_{22} = \frac{y_{22}}{y_{12}}$$

$$a_{11} = -\frac{\Delta h}{h_{21}} \quad a_{12} = -\frac{h_{11}}{h_{21}}$$

$$a_{21} = -\frac{h_{22}}{h_{21}}; \quad a_{22} = -\frac{1}{h_{21}}$$

$$b_{11} = \frac{1}{h_{12}} \quad b_{12} = \frac{h_{11}}{h_{12}}$$

$$b_{21} = \frac{h_{22}}{h_{12}} \quad b_{22} = \frac{\Delta h}{h_{12}}$$

$$a_{11} = \frac{1}{g_{21}} \quad a_{12} = \frac{g_{22}}{g_{21}}$$

$$a_{21} = \frac{g_{11}}{g_{21}} \quad a_{22} = \frac{\Delta g}{g_{21}}$$

$$b_{11} = -\frac{\Delta g}{g_{12}} \quad b_{12} = -\frac{g_{22}}{g_{12}}$$

$$b_{21} = -\frac{g_{11}}{g_{12}} \quad b_{22} = -\frac{1}{g_{12}}$$

$$a_{11} = -\frac{b_{22}}{\Delta b} \quad a_{12} = \frac{b_{12}}{\Delta b}$$

$$a_{21} = -\frac{b_{21}}{\Delta b} \quad a_{22} = \frac{b_{11}}{\Delta b}$$

$$b_{11} = \frac{a_{22}}{\Delta a} \quad b_{12} = \frac{a_{12}}{\Delta a}$$

$$b_{21} = \frac{a_{21}}{\Delta a} \quad b_{22} = \frac{a_{11}}{\Delta a}$$

10.3 Applying two-port parameters

Once you have a set of parameters, you do not need to insert the full subcircuit, but only work with it using the equations, or at most, the equivalent macro model circuit. This one exists for parameters **Z**, **Y**, **H**, and **G** (See Fig. 10.2). When you substitute the two-port by its model, then the circuit becomes another one with only two terminals, and may be analyzed using any of the methods presented in previous chapters. There are also many other useful applications which fall beyond our scope.

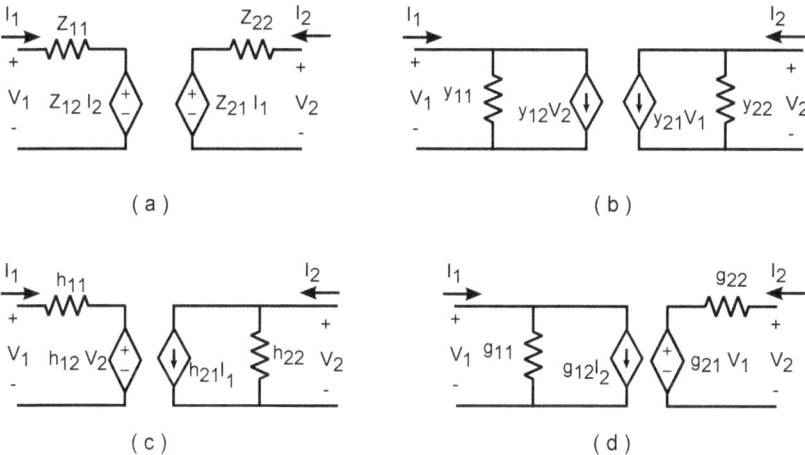

Figure 10.2: Two port equivalent models:(a) from Z-parameters, (b) from Y Parameters, (c) From H parameters, (d) from G parameters

In this section, only two among the many cases are considered: terminated two-ports and, tandem connection.

10.3.1 Terminated two-ports

Fig. 10.3 show two-port networks terminated by a load resistance and a source including a resistance. This source represents the Thevenin or Norton equivalent. For these configurations, we can establish formulas for the different network functions which we may program in our calculator using the two port parameters. There is an example below about finding the formulas. The reader may work any formulas he/she wishes, or else look at tables available in textbooks or internet, and program results. I omit here those tables.

Example 10.2 *Assume the two port is characterized by the hybrid parameters:*

$$V_1 = h_{11}I_1 + h_{12}V_2$$

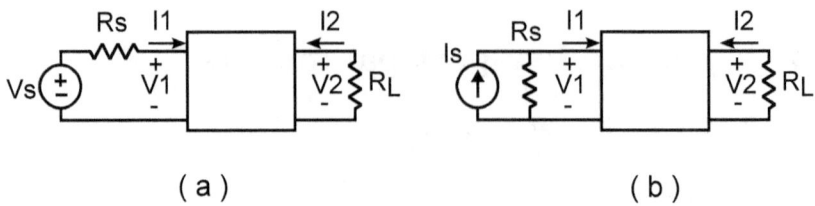

Figure 10.3: Terminated two ports :(a) with voltage source, (b) with current source

and
$$I_2 = h_{21} I_1 + h_{22} V_2$$

Now, let us consider the additional equations that arise from the connection shown in Fig. 10.3:

$$V_2 = -R_L I_2 \quad \text{and} \quad V_1 + R_s I_1 = V_s$$

Using the four equations, we can deduce after some Algebra the following formulas:

$$R_{in} = \frac{V_1}{I_1} = \frac{h_{11} + \Delta h\, R_L}{1 + h_{22}\, R_L}$$

$$I_2 = \frac{h_{21} V_s}{(h_{11} + R_s)(1 + h_{22} R_L) - h_{12} h_{21} R_L}$$

$$R_{th} = \frac{R_s + h_{11}}{\Delta h + h_{22}\, R_s}$$

$$V_{th} = \frac{-h_{21}\, V_s}{\Delta h + h_{22}\, R_s}$$

$$A_{vs} = \frac{V_2}{V_s} = \frac{-h_{21} R_L}{(h_{11} + R_s)(1 + h_{22} R_L) - h_{12} h_{21} R_L}$$

$$A_v = \frac{V_2}{V_1} = \frac{-h_{21} R_L}{h_{11} + \Delta\, R_s}$$

$$A_I = \frac{I_2}{I_1} = \frac{h_{21}}{1 + h_{22}\, R_L}$$

$$G_m = \frac{I_2}{V_1} = \frac{h_{21}}{h_{11} + \Delta h\, R_L}$$

In this table, V_{th} and R_{th} stand for the Thevenin equivalent seen by load R_L. From these formulas we could perfectly define functions. For example a function for the input resistance above could be the following

10.3. APPLYING TWO-PORT PARAMETERS

```
Define RinH(h,rl)= (h[1,1] + det(h)× rl)/(1 + rl × h[2,2])
```

We could also write a program. For example, a partial list of two results in a program could be:

```
terminatedh(h,rs,rl)
:Pgrm
:(h[1,1] + det(h)*rl)/(1 + h[2,2]*rl) → rin
:Disp ''Rin = '' rin
:-h[2,1]/(h[2,2]*rs + det(h))→ D
:Disp ''Vth/Vs = '' D
ETC
```

As illustrated with the example, you may build up your own data base of formulas in your calculator using the different parameters for the network functions. Another alternative, equally valid, is to work with the equations directly and interpret your results. Consider the following example.

Example 10.3 *Assume that in Fig. 10.3(b) Rs = 2.12 kΩ and RL = 1.5 kΩ. Furthermore, assume that the two port is described by the ABCD parameters as*

$$V_1 = 0.125 V_2 - 8 I_2$$

and

$$I_1 = 0.001 V_2 - 1.2 I_2$$

The connection equations are $I_1 + V_1/R_s = I_s$ and $V_2 = -R_L I_2$. Combining all equations with the given values, and taking $I_s = 1$ A for network function calculation, we arrive at:

$$\begin{array}{c} \\ V_1 \\ I_1 \\ Input \\ Out \end{array} \begin{array}{c} V_1 \quad V_2 \quad I_1 \quad I_2 \\ \begin{bmatrix} -1 & 0.125 & 0 & -8 \\ 0 & .001 & -1 & -1.2 \\ \frac{1}{2120} & 0 & 1 & 0 \\ 0 & 1 & 0 & 1500 \end{bmatrix} \end{array} \begin{bmatrix} V_1 \\ V_2 \\ I_1 \\ I_2 \end{bmatrix} = \begin{array}{c} Is \\ \begin{bmatrix} 0 \\ 0 \\ 1 \\ 0 \end{bmatrix} \end{array} \quad (10.4)$$

Solving this equation we arrive at

$$\begin{bmatrix} V_1 \\ V_2 \\ I_1 \\ I_2 \end{bmatrix} = \begin{bmatrix} 70.016 \\ 537.208 \\ 966.97E-3 \\ -358.14E-3 \end{bmatrix} \quad (10.5)$$

Let us store this result in variable **z**. *From the vector we can deduce the following:*

$$\frac{V_2}{I_s} = 537.8\,\Omega(\ \mathtt{z[2,1]}),\quad \frac{-I_2}{I_s} = 0.358\ (\boxed{(\text{-})}\ \mathtt{z[4,1]})\quad R_{in} = \frac{V_1}{I_s} = \mathtt{z[1,1]} = 70.016\,\Omega$$

and so on.

10.3.2 Tandem Connection

A particular structure that in general is simpler to work with two parameters, and easy to program, is the one known as a *cascade* or *tandem* connection of two ports, illustrated in Fig. 10.4. Here, the "output" port 2 of block 1 is connected to the "input" port 1 of the next two-port. The overall system may be considered as a two-port, in which port 1 is the port of the first two-port in the chain, while port 2 is that of the last one in the chain. Notice that $V_{21} = V_{12}$ and $-I_{21} = I_{12}$, which explains why we use the negative of current I_2 in the definition of chain parameters.

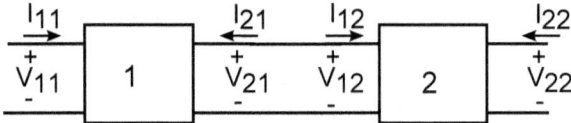

Figure 10.4: Cascade or tandem connection of two ports.

Looking at Fig. 10.4, we see that

$$\begin{bmatrix} V_{11} \\ I_{11} \end{bmatrix} = \mathbf{A_1} \begin{bmatrix} V_{21} \\ -I_{21} \end{bmatrix} = \mathbf{A_1} \begin{bmatrix} V_{12} \\ I_{12} \end{bmatrix} = \mathbf{A_1\,A_2} \begin{bmatrix} V_{12} \\ -I_{22} \end{bmatrix}$$

This means that we can obtain the overall chain parameters using matrix multiplication of the individual port matrices, in the same order in which they are connected.

This principle is applicable in many cases of interest including programming of configurations where two-port configurations in cascade may be identified. Among these configurations we find the popular ladder network, shown in Fig. 10.5. Although more useful in complex networks, it is treated in the same way in resistive circuits.

In this figure, the cascade connection is identified, as a successive tandem placement of the two simple ports called series resistance and parallel resistance, which are shown in Fig. 10.6(a) and (b).

For these two cases, the chain matrices for the series and parallel resistance two ports are, respectively

$$\begin{bmatrix} 1 & R \\ 0 & 1 \end{bmatrix} \tag{10.6}$$

and

10.3. APPLYING TWO-PORT PARAMETERS

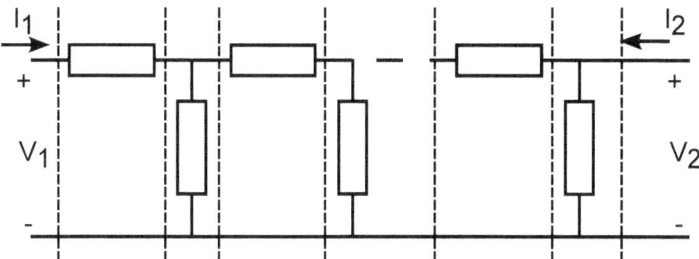

Figure 10.5: Ladder configuration.

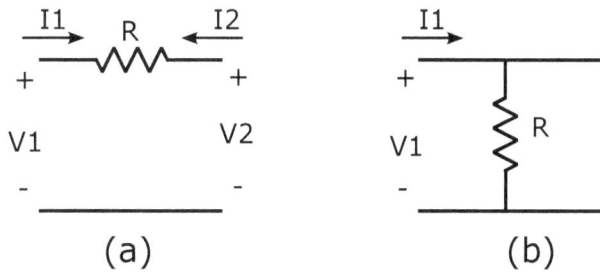

Figure 10.6: Examples of two ports for data base: a) Series R, b) Parallel R

$$\begin{bmatrix} 1 & 0 \\ \frac{1}{R} & 1 \end{bmatrix} \qquad (10.7)$$

There are of course other examples of two ports, like those in in Fig. 10.7, an ideal voltage controlled current source and a voltage amplifier with infinite input resistances. These structures have the following matrices, respectively.

$$\begin{bmatrix} 0 & -\frac{1}{g} \\ 0 & 0 \end{bmatrix} \qquad (10.8)$$

$$\frac{1}{G_f R_o + A} \begin{bmatrix} G_f R_o + 1 & R_o \\ G_f (1 - A) & G_f R_o \end{bmatrix} \qquad (10.9)$$

Notice that the conductance G_f form of the resistance is used. This is because an open circuit has conductance 0, and therefore can be used in programming.

All four cases may be programmed as functions, for example, or else included as subroutines in a program. Examples for function definitions are:

Series R branch: Define serr(r)= [1, r ; 0 , 1]

Parallel R branch: Define parr(r)= [1, 0 ; 1 / r , 1]

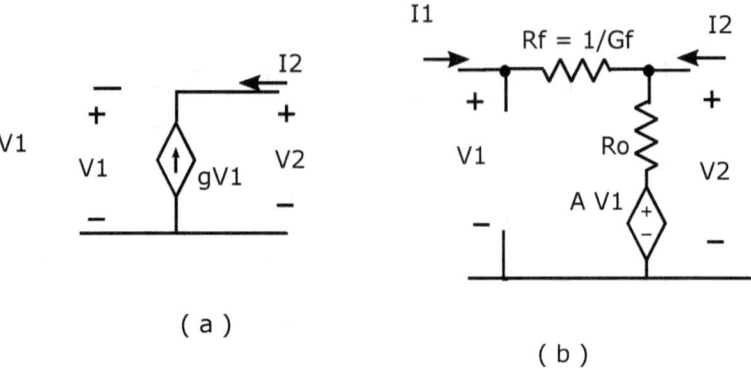

(a)

(b)

Figure 10.7: Other examples of two ports for data base: a) ideal VCCS, b) Voltage amplifier with infinite imput resistance

VCCS: Define vcs2(g)= [0 , (-) 1 / g ; 0 , 0]

Rf_Ro_A two port: (Ro =r, 1/Rf = g with g=0 when Rf = ∞)

Define rga(r,g,a)= (1/(g $\times$ r+a)) $\times$ [(g $\times$ r+1) , r ; g $\times$ (1 - a) , (g $\times$ r)]

Notice that an ideal voltage controlled voltage source is included for the case rga(0,0,a).

Thus, you can create your own database with the two-ports that you use frequently. An example using these functions for the tandem connection follows.

Example 10.4 *Consider the loaded inverting amplifier of Fig. 10.8(a), which has an ideal voltage gain $V_o/V_s = -10$. Assuming an operational amplifier with an input resistance of 200 kΩ, an output resitance of 75 Ω and an open circuit gain of 1000, the equivalent circuit is that of Fig. 10.8(a). Find the actual gain.*

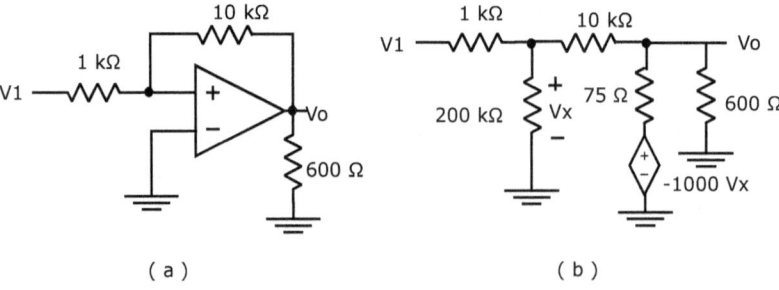

(a)   (b)

Figure 10.8: a) A loaded inverting amplifier, and b) its equivalent circuit .

We can identify the two ports in cascade for this configuration, considering

10.3. APPLYING TWO-PORT PARAMETERS

again the last port as a one terminated with an open circuit. Using then the functions defined before we get the overall chain matrix with the input

serr(1000)*parr(200E3)*rga(75,1/10000, (-) 1000)*parr(600) ENTER
to get

$$\begin{bmatrix} -101.25\text{E-}3 & -82.88\text{E-}3 \\ -100.12\text{E-}6 & -7.88\text{E-}6 \end{bmatrix}$$

From this result, the actual gain is obtained with

$$1/\text{ans}(1)[1,1] \to -9.8764$$

Consider now another example. This one allows us to look at a terminated two port network from another perspective.

Example 10.5 *Take again the terminated two-port of example 10.3 which is redrawn in Fig. 10.9 so that the cascade structure is identified.*

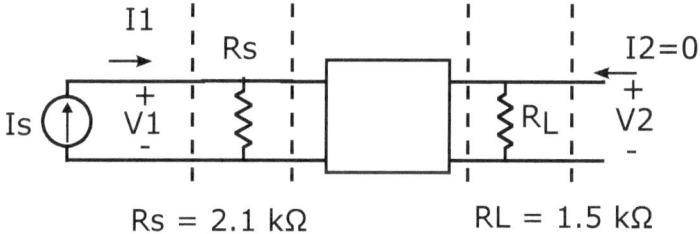

Figure 10.9: Terminated two-port interpreted as cascade of two-ports.

We should be aware of the way the trick works: 1) I_2 of the two-port <u>is not</u> the current in the 1.5 kΩ resistor. For the combined two-port, $I_2 = 0$; (b) the current in the resistor is $I_L = V_2/1500$; and (c) for the combined two-port $I_1 = I_s = 1$.

We can now apply the tandem property:

$$\begin{bmatrix} A & B \\ C & D \end{bmatrix} = \begin{bmatrix} 1 & 0 \\ \frac{1}{2120} & 1 \end{bmatrix} \begin{bmatrix} 0.125 & 8 \\ .001 & 1.2 \end{bmatrix} \begin{bmatrix} 1 & 0 \\ \frac{1}{1500} & 1 \end{bmatrix}$$

$$= \begin{bmatrix} 130.330\text{E-}3 & 1.204\text{E}0 \\ 1.861\text{E-}3 & 1.204\text{E}0 \end{bmatrix}$$

Storing again the new matrix in variable z, we get

$$\frac{V_2}{I_s} = 537.8\ \Omega(\ 1\ /\ \text{z[2,1]}),\quad \frac{I_L}{I_s} = 0.358\ (\ 1/\ \text{z[2,1]}/1500)$$

$$R_{in} = \frac{V_1}{I_1} = \frac{\text{z[1,1]}}{\text{z[2,1]}} = 70.016\ \Omega$$

We see that with the proper interpretation we obtain the same results as in example 10.3, but with less effort.

CHAPTER 11

What's Next

Well, I must finish somewhere. I decided I should limit this introductory book to resistive circuits only, and this is why the book has the subtitle "First part, resistive circuits". Obviously, you may object that there are still too many topics not covered. And you're right. But I certainly covered enough to be sure that you can go ahead by yourself in many ways.

Steady state analysis of circuits is another topic to be covered, which I have left for a companion book on purpose. These circuits may be looked upon in two ways. For applications, we need to look also on the theoretical side, which is why I leave the treatment for the second part. I discuss this point briefly on the next section.

Dealing with linear circuits containing reactive elements is also another topic to be expanded in a later book. In section 11.2, I show with an example that this area is already advanced with the methods already developed here.

In short, consider this chapter an appetizer for going further into circuit analysis using the calculator. Don't forget, however, that the calculator is not the teacher, only the tool.

11.1 Complex circuits in steady state domain

As far as linear circuits concern, all the analyses done here are directly applicable to complex impedances and admittances. The TI-89 allows handling of complex numbers in lists and matrices, so you should find no problem in applying the methods and programs mentioned here. Steady state analysis includes however other topics, and the companion book intends to apply the calculator to help you learn those topics as well.

Let me introduce briefly some tools for you to use in successful application of the calculator.

11.1. COMPLEX CIRCUITS IN STEADY STATE DOMAIN

11.1.1 Elementary notation and display for complex numbers

A complex number z is in rectangular form if it is written as $z = a + bi$, where $i = \sqrt{-1}$ is the imaginary unit. The real and imaginary parts of the complex number z are defined as $\text{Re}(z) = a$, $\text{Im}(z) = b$, respectively. **In circuits we use j for the imaginary unit, but calculators use i.** In this book, we adhere the the j notation for the circuits but use i for the calculator keys.

On the other hand, in polar form, z is written as $z = r\angle\phi = re^{i\phi}$. Here, the absolute value of z is $\text{abs}(z) = |z| = r$ and the angle or argument of z is $\arg(z) = \text{angle}(z) = \phi$. We always assume $r \geq 0$. $\angle\phi$ is the shorthand notation for the Euler identity

$$e^{i\phi} = \cos\phi + i\sin\phi \tag{11.1}$$

usually used in engineering textbooks.

Graphically, the complex number z may be represented by a vector in a two dimensional Cartesian plane, called complex plane. The real part is represented in the x-axis, and the imaginary part in the y-axis. The absolute and angle values are then the polar coordinates of the vector. Very often in circuits we try to limit the angle to the values $-180° \leq \phi \leq 180°$. Fig. 11.1 illustrates some examples for complex numbers.

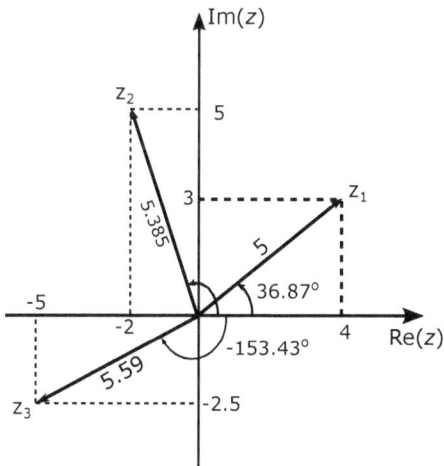

Figure 11.1: Graphical representation of $z_1 = 4 + 3i = 5\angle 36.87°$, $z_2 = -2 + 5i = 5.385\angle 111.80°$ and $z_3 = -5 - 2i = 5.59\angle -153.43°$ in the complex plane.

11.1.2 Complex numbers and calculator

To work with complex numbers in your calculator prepare your settings accordingly. Do not work in Real mode, because your calculator returns an error for a complex-number result, except if you enter your expression using i, the imaginary unit.

To enter the imaginary unit i: press 2nd CATALOG
Settings for display:

- In complex settings, select either rectangular mode $(a + bi)$ or complex polar mode

- Must choose angle settings (radians or degrees) if polar mode was selected

Notice that the input may be in either rectangular or polar mode, but the result will always be displayed according to the selected setting.

Now, since the polar input and display is different depending on the calculator family, let us mention the differences

Lists and Matrices:

The TI-89 afamily accepts complex numbers as usual in lists and matrices. Hence, there is no problem using these tools in all formulas and procedures.

11.1.3 An Example

Let us see one example using transformations and applying lists. The solution of the problem could also be done using nodal or loop equations I encourage the reader to try them.

Example 11.1 *Consider example 5.20 on page 82. In this example, the current I was found using transformations and lists. Let us now work a "similar" problem, with the same topology, asking for a current in the same branch, and following the same steps, but in the complex plane.*

Consider the circuit in Fig. 11.2(a); the solution was obtained then reducing the circuit to that shown in (b) by first applying source translation and then Thevenin equivalent calculated simultaneously for both extremes. I recommend the reader to see again the original example where the steps are illustrated graphically. Then, draw the steps for this examle.

Now, I will follow exactly the same steps as in example 5.20, except for the fact that I am using complex numbers in the lists, and for the voltage. Settings for the display are polar and degrees for the display, with two decimal figures, **except for current where ENG mode is used** *since the current magnitude in mA is not displayed with the original settings. Notice that polar inputs are in parenthesis.*

The table that follows is similar to the one used in the original example 5.20.

11.1. COMPLEX CIRCUITS IN STEADY STATE DOMAIN

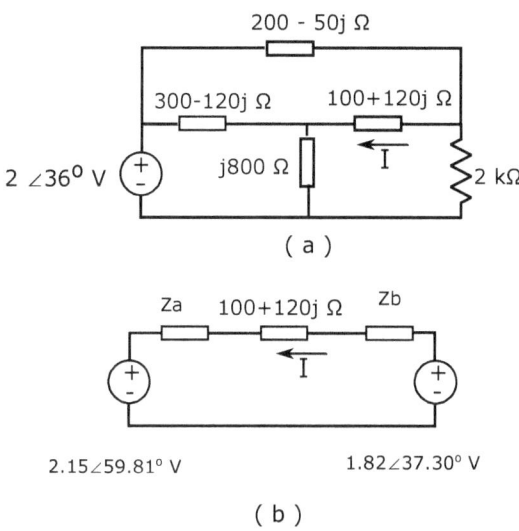

Figure 11.2: Reduction for a complex bridged-T. Za = (347.79∠2.00) Ω, Zb = (187.37∠-12.73) Ω

Entry: {300-120i, 200-50i} STO▶ Z1
In Stack: {(323.11∠ -21.80) (206.16∠-14.04)}
Entry: {800i, 2000} STO▶ Z2
In Stack {(800.∠ 90.) 2000.}
Entry: (2∠36) × Z2 ÷ (Z1 + Z2) STO▶ Vt
In Stack: {(2.15∠59.81) (1.82∠37.30)}
Meaning: Vt = {Va, Vb}
Z1 × Z2 ÷ (Z1 + Z2) STO▶ Zt
In Stack {(347.79∠2.00) (187.37∠ -12.73)}
Meaning: {Zta, Ztb}
(Vt[2]-Vt[1]]) ÷ (100 + 120i + sum(Zt))
In Stack (1.32E-3∠-72.63)
Meaning: Current**

The complex current that results is hence $1.32\angle -72.63°$ *mA. In rectangular form, you may convert to get* $(0.394 - 1.26\,i)$ *mA. Remember though that voltages and currents are usually preferred in polar form.*

11.2 Short reference to time domain circuits

To the companion book I have also left the introductory topic to time domain analysis. As an appetizer, consider the conceptual example shown in Fig. 11.3 for a first order circuit. Inset (a) shows the general configuration. To find the current and voltage at the capacitance C, we calculate the Thevenin equivalent, for example using a current source (Section §9.1), as seen in inset (b). We arrive then at the simple RC circuit of Fig. 11.3(c), in which we find voltage $V_c(t)$ and current $i_c(t)$.

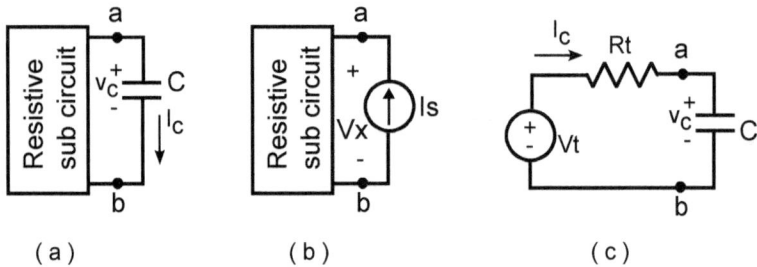

(a)   (b)   (c)

Figure 11.3: A first RC circuit: (a) Original configuration; (b) Ready to find the equivalent seen by capacitor; (c) Circuit to find current and voltage at capacitor.

Once you have the current $i_c(t)$, the substitution theorem (Section 4.1) states that the capacitor may be substitute by a current source, which is the case of (b) if you make $I_s = -i_c(t)$. This means that you already have all the information, because of superposition theorem, to find out all the voltages and currents in your circuit if you desire to do so. Let us work one final example.

Example 11.2 *Let us take the circuit from example 9.1 on page 169, terminated with a capacitance of 12 µF as shown in Fig. 11.4. Let us assume that the initial condition at the capacitance is $V_C(0) = 1$ V*

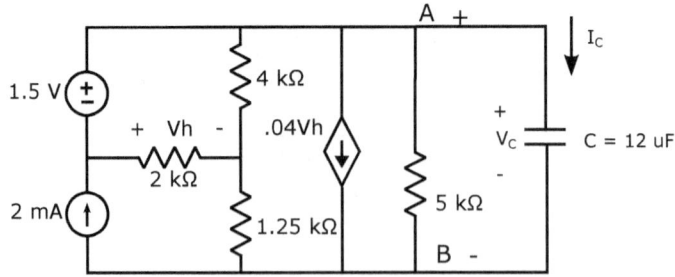

Figure 11.4: A first order circuit to illustrate the procedure

Since the sources in the circuit are constant, the theory of first order circuits

11.2. SHORT REFERENCE TO TIME DOMAIN CIRCUITS

tells us that the voltage $V_C(t)$ and current $I_C(t)$ at the capacitance for $t \geq 0$ are given by

$$V_C(t) = V_{th} + (V_C(0) - V_{th})e^{-t/\tau} \tag{11.2}$$

and

$$I_C(t) = \frac{V_{th} - V_C(0)}{R_{th}} e^{-t/\tau} \tag{11.3}$$

where the we have the time constant $\tau = R_{th} C$.

Substituting the capacitance by a current source as illustrated in Fig. 11.3(b) takes us to Fig. 9.5(a) on page 170. For this circuit, the nodal equations were set in the example and solved. Let us store the solution in variable B From the example we have

$$\text{yn}^{-1} * \text{in} \rightarrow \begin{bmatrix} 526.74\text{E-}3 & 47.098\text{E}0 \\ 496.818\text{E-}3 & 22.789\text{E}0 \\ 2.0267\text{E}0 & 47.098\text{E}0 \end{bmatrix} \boxed{\text{STO}} \quad \text{B} \tag{11.4}$$

From this result we get (I use TI-89):

τ : b[3,2] × 12 $\boxed{\text{EE}}$ $\boxed{(-)}$ 6 $\rightarrow$ 565.18E-6
$1/\tau$: $\boxed{\wedge}$ $\boxed{(-)}$ 1 $\rightarrow$ 1.7694E3
$V_C(0) - V_{th}$: 1 - b[3,1] $\rightarrow$ -1.0267
Coefficient in I_C : $\boxed{(-)}$ ANS/b[3,2] $\rightarrow$ 21.8E-3

Therefore, for $t \geq 0$,

$$V_C(t) = 2.0267 - 1.0267 e^{-1769.4 t} \quad \text{V}$$

and

$$I_C(t) = 21.8 e^{-1769.4 t} \quad \text{mA}$$

Furthermore, from superposition (11.4) provides us the expression for all node potentials, since source $Ix = -I_C$. We can either work individually or perform the following operation:

$$\mathbf{B} \boxed{\times} \begin{bmatrix} 1 & 0 \\ 0 & c \end{bmatrix} = \begin{bmatrix} 526.74E-3 & -1.0267 \\ 486.81E-3 & -496.81E-3 \\ 2.0267 & -1.0267 \end{bmatrix}$$

Observe that the third row are the coefficients for V_C, as expected. The first row tells us that $V_1 = 0.527 - 1.027 e^{-1769.4 t}$, and similarly for the other row.

Once we have the potentials, we can work any other variable such as current or power.

11.3 Final remarks

As the examples above show, with the methods presented in this book you already have the fundamentals to continue with more applications. Of course, there are so many more things and methods to discover. So, the answer to "what's next?" might be, "you may go on, you have the tools now."

But never forget: the calculator is your tool, not your teacher. It's up to you to make good use of it.

CHAPTER 12

References

The references below provide some sources I have used either to present the theoretical methods, learn about calculators for tricks that are not included in the guidebooks, and fundamentals to justify the methods presented here for using the calculator. Although some of the techniques presented here for exploiting the calculator were independently developed by myself, I cannot claim being the first one using them, since sometimes I discovered later that the method had already been introduced by someone else.

In the mentioned references there are other methods that may be implemented in the calculator. The interested reader may consult a reference to find out more of these, and have fun in addition to learning.

There is a lot of room for improvement and for introduction of more methods. Up to you!

Bibliography

[Ayres1962] Ayres Jr, Frank, *Matrices*, Schaum Publishing Co., New York, 1962.

[Balabanian69] Balabanian, Norman and Bickart, Theodore A and Seshu, Sundaram, *Electrical network theory*, John Wiley & Sons, Inc, New York, 1969

[Boite1976] Boite, René, and Neirynck, Jacques. *Thorie des rseaux de Kirchhoff*. Georgi, 1976.

[Budak87] Budak, Aran, *Circuit Theory Fundamentals and Applications*, Prentice-Hall, Englewood Cliffs, NJ, 1987.

[Cacovean] Cacovean, Andrew, *Making Custom Menus*, available at http://www.ibiblio.org/technicalc/examples/en/html/examples_index/custom-menus.html

[Chua69]	Chua, Leon O, *Introduction to nonlinear network theory*, McGraw-Hill, New York, 1969.
[Edward05]	Edwards, Constance C, *TI-89 Graphing Calculator for Dummies*, John Wiley & Sons, Inc, New York, 2005
[Karni1966]	Karni, Shlomo, *Network Theory: Analysis and Synthesis*, Allyn and Bacon, 1966.
[Kiss68]	Kiss, W.F., Gilson, R.A, *On the formulation of the indefinite matrix*, IEEE J. of Solid-State Circuits, vol. 3, No. 3, pp. 307-308, 1968
[Moschytz1974]	Moschytz, George, *Linear Integrated Networks: Fundamentals*, Van Nostrand-Reinhold, New York, 1974.
[Palomera06]	Palomera-García, Rogelio. "Multipole and Multiport Analysis", in *Wiley Encyclopedia of Electrical and Electronics Engineering online* (John G. Webster, Ed.), John Wiley & Sons, Inc, (2001, updated 2006), URL: http://dx.doi.org/10.1002/047134608X.W2506
[TI89]	Texas Instruments, *TI-89 Titanium / Voyage^{TM} 200 Guidebook*, to download go to https://education.ti.com and select "Downloads" menu.
[Vlach83]	Vlach, J., and Singhal, K. *Computer methods for circuit analysis and design*. Springer Science & Business Media, 1983
[Voltmer99]	Voltmer, David R and Yoder, Mark A, *Electrical Engineering Applications with the TI-89*, Texas Instruments, Inc., 1969

www.ingramcontent.com/pod-product-compliance
Lightning Source LLC
Chambersburg PA
CBHW070231190526
45169CB00001B/155